T0173570

BRIAN
TRUBSHAW

This book is dedicated to all those involved in the design and development of civil and military aircraft and their associated systems and weapons. It is especially dedicated to those many test pilots and flight test engineers who have given their lives in the furtherance of aviation.

BRIAN TRUBSHAW
TEST PILOT

BRIAN TRUBSHAW

WITH

SALLY EDMONDSON

FOREWORD BY HRH
THE DUKE OF EDINBURGH

First published in 1998 by Sutton Publishing
This new paperback edition first published in 2006

Reprinted in 2016 by
The History Press
The Mill, Brimscombe Port
Stroud, Gloucestershire, GL5 2QG
www.thehistorypress.co.uk

Reprinted in 2017

British Library Cataloguing in Publication Data
A catalogue record for this book is available from the British Library

ISBN978 0 7509 4494 6

Typeset in 10.5/12.5pt Photina.
Typesetting and origination by
Sutton Publishing Limited.
Printed in Great Britain by TJ Books Ltd, Padstow, Cornwall

Contents

List of Plates

Foreword
by
HRH The Duke of Edinburgh

BUCKINGHAM PALACE.

Brian Trubshaw must be one of the last of a line of what were perceived as the intrepid test pilots. Test pilots certainly still exist but, as the aviation industry has become more established, the people who test-fly new aircraft have rather lost their hero status and become part of the whole production system.

This account of his time as a test pilot covers the period of the most rapid development in aircraft design. It was a time when new names of test pilots were famous in every household with an interest in aviation. Less than forty years saw the transition from small and comparatively slow unpressurized piston-engined passenger planes to the supersonic Concorde. Each step was a leap forward, and each one posed a formidable challenge to the pilots who had to prove that they could fly. This book chronicles the human side of a technological revolution.

Acknowledgements

I have received a great deal of encouragement from my wife Yvonne and from my step-daughter Sally Edmondson to write this book which is based on my own memories, my flying log books and my own test notes on Concorde. Sally has helped me put the book together, as well as carrying out all the typing and preparation, and I am most grateful to her.

I have received particular help from Dr Norman Barfield concerning all the aircraft emerging from Weybridge. I am also grateful to Howard Berry, Shaun Greene and Mike Fish of British Aerospace Airbus, Filton for some Concorde reminders. I am most grateful to Howard Berry, on behalf of British Aerospace, for allowing me to use the illustrations and photographs that are included in this book and to Shaun Greene for his help in their preparation. Unless otherwise stated, all photographs are reproduced with the kind permission of British Aerospace.

I also wish to thank Ken Smart, Head of Air Accident Investigation Branch for granting me permission to reproduce the accident report on the crash of the prototype BAC1-11.

The chapters on Concorde were the easiest to put together because my involvement in the programme was the most recent. I have a considerable collection of almost every book that has been written on Concorde which includes: Geoffrey Knight, *Concorde*; Brian Calvert, *Flying Concorde*; Christoper Orlebar, *The Concorde Story*; T.E. Blackall, *Concorde*; R.E. Gillman, *Croydon to Concorde*; André Turcat, *Concorde*; John Costello and Terry Hughes, *Concorde*; and Reginald Turnhill,

Celebrating Concorde. I have also made reference to Ken Owen's latest book, *Concorde and the Americas*, as well as his earlier book, *Concorde, – New Shape in the Sky.*

Both Sally and I would also like to thank Lynn Cowley, Pippa Houghton and Carmel O'Connor for their IT support.

It has not been possible to mention, by name, all of the many test pilots I was associated with, all the design engineers, flight test staff, maintenance staff, inspection staff, test service organization, air traffic control and administrative personnel. I thank you all for your friendship, skill and help.

E.B.T.
August 1998

Introduction

I have to admit that I was decidedly dry in the mouth when I lifted Concorde 002 off the ground for the first time on 9 April 1969. The short flight was from Filton, Bristol to RAF Fairford in Gloucestershire, which would be our flight test base for the next seven years. Not only had a long, hard process finally reached fruition for all those involved in the production of this magnificent aircraft, but, for me, the occasion undoubtedly marked the pinnacle of the nineteen years of test flying that I had completed so far.

I was tremendously excited at the prospect of flying 002 for the first time and I felt deeply proud that I had been chosen to do it. Many test pilots would have given almost anything to be in my shoes and I well appreciated how lucky I was. I was also very aware of the responsibility that lay in my hands without having to worry about the publicity, which rammed home any shortfall that I may have had. First flights are easier to conduct without the huge razzmatazz that surrounded Concorde.

Concorde 002 looked extremely smart from the outside but inside what was supposed to be the passenger cabin was so jammed with flight test instrumentation for recording flight data it was difficult to appreciate that it would become the first supersonic civil airliner.

All the roads around both Filton and Fairford were packed with onlookers. At Filton, all the BAC and Rolls-Royce employees gathered on each side of the airfield to observe the events of this great day. A day which meant a great deal to so many of them because of their own individual contributions. Milling around the Flight Operations Building at Filton was a bevy of VIPs which included Sir George Edwards, Chairman of BAC, General Henri Ziegler, Chairman of Aerospatiale, Mr Anthony Wedgwood Benn, the Minister of Aviation, Sir James Hamilton, Director-General Concorde and other Government officials, as well as the late Sheila Scott, world famous for flying solo around the world, and Aerospatiale's Chief Test Pilot, André Turcat.

It was an awesome experience for myself and my crew. Deep down we were full of emotion but there was little sign of it as we settled to our task. This flight was different to anything that any of us had encountered before. The eyes of millions of people all over the world were focused on us, as they had been on our French counterparts six weeks earlier when 001 flew from Toulouse. As a crew we were happy and determined. We had all reached this point from different directions and it had not happened overnight for any of us. It had been a long and exacting experience.

I had flown many times with John Cochrane, the co-pilot on 002's first flight, initially when he was still in the Royal Air Force flying V bombers at Boscombe Down as a test pilot before he joined BAC at Weybridge on the VC10 programme. I had not, however, flown with the other members of the first flight crew before: Flight Engineer Brian Watts, Senior Flight Test Observer John Allan, Flight Test Observers Mike Addley and Peter Holding, or Navigator Roy Lockhart. However, we had all been together on the simulator many times and had carried out the whole engine running programme as a complete crew and I had total confidence in them all. We were a highly professional and well-organized crew.

The two previous days had been very frustrating because each time I attempted to complete the fast taxi tests an instrument failure flag on the Captain's airspeed indicator kept appearing. I needed to carry out these fast runs in order to check the correct response from flight control inputs. Frantic efforts to cure the fault by the test engineers, assisted by the design office, failed to resolve the problem the first and second time, but eventually they were confident they had sorted it out. I decided that, if this was the case, I would keep going and get airborne. I could see from the approaching upper cloud that the wonderful weather that we had been enjoying for several days was going to break as a warm front headed in from the west. If we were going to go, now was the time. The instrument failure flag did not reappear and Concorde 002 was airborne at last. The crew were fairly quiet so I reassured them that everything was going well – we were on our way to what can justifiably be regarded as one of the greatest flight test programmes of all time. More importantly, from that moment onwards, our total involvement in the Concorde flight test programme began in earnest.

The day before telegrams had poured in to wish us well, including one from an old school friend of mine, which read 'Good luck tomorrow. Make sure you don't break your bloody neck!'

ONE

The Early Days

A single event in 1934, when I was ten years old, had a most profound effect on my future. This was the occasion when the Prince of Wales's aircraft landed on the beach at Pembrey, Carmarthenshire (Dyfed) as a prelude to the prince's (later King Edward VIII) opening of the Royal Welsh Agricultural Show later that year.

It was my first sight of an aeroplane and as it circled overhead I ran at great speed across the Ashburnham Golf Course, which lay between our house and the beach, so that I could observe it landing. I can still recall my excitement as I ran after the aircraft with a few other enthusiastic fools. Although, when the aircraft turned around to face us after landing, it was something of a surprise to see the pilot shaking his fist at us. I later found out that the pilot was Flight Lieutenant 'Mouse' Fielden (Air Vice-Marshal Sir E.H. Fielden GCVO, CB, DFC, AFC) who had been transferred to the Reserve of Air Force Officers.

My family lived in South Wales because my father was in the tin plate side of the steel business at the Western Tin Plate Works, Llanelli. The Trubshaws originally hailed from Little Heywood in Staffordshire, where the family home was known as Trubshaw Hall. My grandfather, Ernest Trubshaw, a formidable character by all accounts, married Lucy Smith of Barrow whose father, an industrialist, sent his new son-in-law to Llanelli to run the family owned Western Works. My grandparents had five children, my father, Harold being the eldest of Charles, Ralph, Gwendoline and Katharine. I think Charles, who was known by the family as 'Barley Boy', was my grandmother's favourite. Apparently she cried so much when packing his trunk at the end of each school holiday, that all his clothes had to be aired on arrival at Uppingham School.

My mother was christened Lumley Victoria Gertrude Carter, but was known throughout her life as 'Queenie'. She met my father during the First World War when he was serving in the Royal Engineers based at Conway. The Carters came from near Caernarvon, North Wales and her sister and four brothers all lived in that area, or in Cheshire, for many years except for Guy who was in the RAF. As a child we visited my Uncle Hamil in Cheshire and regularly spent holidays in Anglesey.

My grandfather's home at Ael-y-Bryn is now a hotel called The Diplomat, but my mother, father, sister Brenda and myself lived a few miles away at Caerdelyn. My grandfather died before I was born, but I remember my grandmother very well. Every week we would take a bus from Caerdelyn to Ael-y-Bryn to visit my grandmother for afternoon tea, which was quite a ritual for someone aged five. My grandmother's face and hair had been very badly burnt when trying to blow out a methylated heater under a teapot. As a result of this accident she used to wear a wig, which became a bit lop-sided from time to time.

We moved to Pembrey in 1927 and Caerdelyn was subsequently bought by my Aunt Gwendoline, who became Dame Gwendoline at the end of the Second World War in recognition of her work in the WVS.

We lived at Pembrey until 1951 when my father died. Soon after my father's death, my mother went through a dreadful ordeal when an intruder dressed from head to toe in black entered her room one night. She lectured him on honour and gave him the three odd pounds in her handbag. She then led him down the stairs and out through the front door. She opened the door ten minutes later to let her spaniel out only to find him still standing there. The intruder turned out to be a local youth my mother had known since he was a small boy. He received a sentence of five years for his crime, but my mother felt sure that she did not want to be around when he came out, so she moved to Beaumaris in Anglesey where one of her brothers lived.

The house at Pembrey was on the edge of the fourth tee of Ashburnham Golf Course, so it is no surprise that we all played golf. My father's handicap was scratch, while his brothers Charles and Ralph were plus one and plus two respectively, and consequently the Trubshaw name is very prominent in the Ashburnham Club House. I never quite aspired to my father's or his brothers' handicaps and did not better a handicap of nine. I did win the Welsh Junior Championship one year when I beat my great rival Gordon Hopkins, although this achievement was rather short-lived as I was beaten the following year by a Miss Roberts. Golf has remained one of my hobbies during adult life, but I never practised enough to fulfil my true potential, in spite of being a very long hitter.

The road across Ashburnham was a dirt track and not a public thoroughfare. Therefore it was ideal for learning to drive, which I was able to do quite easily from the age of twelve. One of my closest friends, John Mansel-Lewis, whose home was also on the golf course, bought a Morgan three-wheeler. The battery was almost always flat, so a push start was always required. This is where my strength was useful as I got a turn at the wheel in exchange. John Mansel-Lewis, whose cousin Sir David is Lord-Lieutenant of Dyfed, later served in the Battle of Britain on Spitfires but sadly lost his life in Singapore.

From an early age all sports were of great importance to me. My father used to take me to watch Llanelli play Rugby football throughout the winter season, but I did not go to a Rugby international until I was eleven years old, and by then at school. All these activities made for a very pleasant life, but then came the day when I was rudely awaked.

TWO

School

In 1932, at the age of eight, I was sent away to a preparatory school called Twyford near Winchester. My earlier education had been a mixture of my sister Brenda's governess and the local curate. My mother took me by train via Paddington and Waterloo to start my first term at Twyford. I was dressed accordingly in the Eton Collar, which was worn every day together with a straw boater. I managed to cope quite well with my new surroundings until my mother left. It suddenly sank in that, for the first time in my life, I was away from home, alone in a completely new environment. The heavens opened and although the tears took a day or two to subside I began to settle into a life I eventually came to thoroughly enjoy.

Twyford was run by an old Wykehamist called H.C. McDonnell (known as 'Doni') who just pipped my father to a golf Blue at Cambridge by coming back for an extra year at university. I soon learned that Twyford was rather a tough place: a cold bath started the day and beatings by 'Doni' were frequent and hard. It was a very stoical establishment where items like sweets and chocolate only featured once a week. 'Doni' had a spinster sister whose main duties at Twyford were to read to us on Sunday evenings and to hand out one sweet to each boy on Thursdays. I took a particular dislike to the lavatories, with the exception of a urinal which flushed itself every few seconds to impress visiting parents. The loos themselves were 'Thunder Boxes' without water. They were, thankfully, emptied each day, their foul contents ferried away in a cart drawn by an unfortunate old horse, who wore boots so that its hooves did not mark the playing fields during the course of this duty or when carrying out its other, less odorous, task – pulling the gang mower.

'Doni' was also a cricket Blue and a fine athlete in his day and games were extremely important within the school. We played soccer, not Rugby, because Winchester played soccer and we played Winchester football for the same reason. I soon made my mark in all games although I was disappointed not to play Rugby as my father and his brothers had all been very good Rugby players. My father's youngest brother Ralph played for Cambridge University, Llanelli and had a trial for Wales, but I had to wait until I joined the RAF before I really got to grips with this great game.

Twyford had extremely high scholastic and games standards, with the aim that its students would go on to Winchester. I was destined to go there in the autumn term of 1937 and was due to take the entrance examination in the summer when the unexpected happened. A vacancy, due to examination failure, came up in 'Doni's' old house ('A' Furleys) a term early, so I was given twenty-four hours' notice to take the exam. This was a great opportunity as I had always wanted to go to Furleys and not to Trants ('E' House), where it was originally planned I go because of the availability of places. I had heard about the relative virtues of these houses from boys who had already gone there. It was with great fear and trepidation that I faced the ordeal. However, by good fortune I managed to pass, but this was not the end of it. 'Doni' took a good look at my answers and was not too pleased. 'You have passed, but you should have done better! I would have expected you to have got a scholarship to anywhere else. Now bend down!' Six Jimi Storkers across my backside followed. I never did understand his philosophy but, as I took the beating I was still feeling rather pleased with my achievement. The beatings at Twyford were frequent and today's society would probably consider that we were all victims of physical abuse, but I do not believe we were any the worse for it. So, after six of the best, I ended my sojourn at Twyford, but not before 'Doni' provided the fullest explanation of the facts of life, thus completing one's preparation for the future.

After I left Twyford, the number of students diminished from the normal 65–70 boys. This was mainly due to the fact that 'Doni''s regime was seen as much too tough for many prospective parents, but the school soon regained its reputation for the highest standards thanks to the Wickham family. 'Doni's' successor, Bob Wickham was more human, his charm and personality giving parents great confidence. Bob also introduced a number of modernizing policies within the school and the 'Thunder Boxes', the foul loos which alone had been successful in putting many parents off sending their sons to Twyford, were removed.

I arrived at Chernocke House (Furley's), Winchester, in May 1937. My mother deposited me there for my first term after having tea in the Norman Meade Hotel, Winchester. Also taking tea there with his mother was another new boy, Richard Rennick (Richard Grant-Rennick). Richard and I became lifelong friends and now live next door to each other. Richard, a Captain in the 6th Airborne (Armd) Recce Regiment, was severely wounded on 22 August 1944. Ironically, almost thirty years later, my step-daughter Sally would arrive at Hatherop Castle near Cirencester for her first term at boarding school where she met another new girl, Richard's youngest daughter, Rhona, who became one of her closest friends.

I went on to enjoy five happy years at Winchester, except perhaps for part of my final year when I was ready to go to war and had already lost a number of my friends, who gave their lives at such a tender age for the good of their country, while many others, like Richard Grant-Rennick, were badly wounded. Richard and I have a great mutual friend in Dick Edward-Collins, another contemporary at school who was also badly wounded in 1944 while serving with the Grenadier Guards. These brave men, both of whom are now seventy-five, decided to put two fingers up to those who said that they would never live beyond twenty and twenty-five. Both, like many others, are to be admired for their great courage.

I had the extreme good fortune to have one of the greatest housemasters of all time, Harry Altham. Harry was a remarkable man who combined all the attributes of a sportsman and scholar. Cricket was perhaps his greatest love and his contribution as a player, coach, historian and later as Treasurer and finally President of the MCC, was enormous. Above all he was a lifelong friend to the hundreds of boys that passed through his house and you knew you'd made the grade at your final goodbye when he would normally say, 'Call me Harry!' His wife, Alison, was a wonderful asset and managed to keep us well fed throughout the difficult war years. Harry's son, Dick and myself played cricket together in the RAF and I still see one of his two daughters, Meg, who married Podge Brodhurst, a junior master when I arrived at Winchester and later a housemaster himself.

Harry Altham taught me an enormous amount about life in a wider sense, especially understanding problems, dealing with people, having standards to live up to and leadership qualities, and I owe him a great deal. He talked in machine-gun sentences while his prose was of an exceptional standard. An example of this machine-gun way of talking is well illustrated by his response to a particular comment made by myself about one of the rather more flamboyant fathers. Harry replied: 'Flash Alf!'

The first undertaking on arrival at Winchester was to learn the 'Winchester Notions', which were a language in themselves: a ball was a 'pill', working was 'mugging', in class was 'up to books', a pretty girl was a 'cud girl', a Rugby scrum was a 'hot'. The school song is 'Home Sweet Home' and sung in Latin standing on chairs. You had to know the names of all the houses, their housemasters, their colours, other appointments such as Head of School, Captain of Cricket and so on. There were a variety of other rules and practices to be learnt and after two weeks an oral examination took place. A second attempt was permitted, but failure resulted in a beating from the Head of the House, who always carried out this task except on the rare occasions when the Headmaster performed it.

Once this preliminary phase was over, life settled into a well-established pattern. The companionship of others and the sporting opportunities made life very tolerable and few were unhappy. I worked and played very hard and was successful in both areas. I came top in my form for the first two years, 'raising books' as it was called, and I made my presence known on the sports field, becoming a member of the colts cricket and soccer sides very quickly.

A bicycle accident cost me a place in the cricket XI in 1940, but I made up for it in 1941 when we beat Eton for the first time for a number of years. We had a good side, which included Philip Whitcombe and Hubert Doggart who both subsequently played for England. Although we did not start too well, 31 for 2 and 66 for 3, we bowled the Eton side out for 190. E.W. Stanton, the well-known cricket commentator and correspondent wrote, 'Trubshaw was then joined by Doggart and when the score rose to 140 before Trubshaw left, the end was in sight. The calling for runs by these two, as well as their batting, was admirable.' In 1942 when I was Captain of Lords, the Eton XI captain was E.W. (Dwin) Bramall, now Lord Bramall of considerable fame. On this occasion Winchester was heavily defeated but a friendship began between myself and Dwin that still exists today.

I made one very good friend in my house called Anthony Bibby and I used to spend part of the school holidays at his home in Surrey. I enjoyed my visits there and his parents, Leslie and Peggy, provided me with many different opportunites. I practically lived with them when I joined

Vickers-Armstrongs in 1950 as their home was only 12 miles from where I worked at Wisley. When Anthony died from polio prematurely at the age of twenty-seven, I tried to repay his parents as much as I could. Now they have both gone, but I still have the lasting friendship of their other son Paul and his wife Janette, their daughter Joanna and her husband Duncan Macleod, as well as their respective families, two of which are my godchildren. I have much to be very grateful for and I look back on many good times, particularly when I had the chance to shoot, stalk and fish, activities I would not have tried without them.

After one year at Winchester, every boy had to join the Officers Training Corps commanded by one of the housemasters of the time, Major Jack Parr. Jack Parr took very unkindly to the re-naming of the OTC to the Junior Training Corps and wrote on the school notice board 'As long as I have the honour to command this Corps, its purpose will be to train officers'. I had one major contretemps with Jack Parr on a field day when he saw Anthony Bibby and myself bicycling on the local golf course. 'What the hell are you doing?' he shouted. I replied, 'We are the balloon barrage Sir!' Jack Parr immediately thought I was taking the mickey, but he apologized profusely when he found out that it was true.

An Air Section was established within the OTC and when it had a field day, the Air Section went to the RAF station at Andover and was given lectures and air experience in Blenheim bombers and Anson trainers. This arrangement was subsequently changed to take place at the Royal Naval Air station at Worthy Down using Fleet Air Arm Sharks, forerunner to the famous Swordfish and Skuas. Unfortunately, during these formative flying encounters, I suffered from acute air sickness, but, after the first disaster, I was always wise enough to take my sponge bag with me whenever I was airborne. My mother inevitably became rather fed up with buying a new sponge bag at the beginning of each school term.

I experienced my first flying accident at Worthy Down when, while we were waiting our turn to take off, another Shark landed on top of the one I was firmly strapped into for no good reason other than finger trouble by the pilot.

Our Liaison Officer was a most delightful and charming man, Lieutenant Olivier (later to become Sir Laurence Olivier). We got on very well with him and caused him some amusement when we equipped ourselves with several bottles of wine to consume with our school-supplied sandwiches. Unfortunately, a similar view was not taken by the master in charge of our outing, Ian McIntosh, who gave me a real roasting when we got back to Winchester. I can still see the shaking finger of this rather short-fused Scotsman, whom I eventually came to like very much indeed. As soon as the Air Training Corps was established nationally, Winchester formed its own Flight and I became a Flight Sergeant.

After three years, I had gone a long way up the school, the war was well under way and life was beginning to change. I became a school prefect for two years and head of my house for my last year. I spent more time playing games than working 'up to books'. All the younger masters had gone to war and had been replaced by some real 'old dodderers'. I was in a physics class being taught by one of them when it became necessary to black out the windows. When the black-out was over, the master in question must have been somewhat surprised to discover we had all left.

By 1940, the effects of the Second World War were beginning to be felt. I had already decided to make the Royal Air Force my career and I had been encouraged to do this by one of my mother's brothers, Air Commodore Guy Carter, who subsequently commanded the fighters in the Western Desert. Guy was killed in Yugoslavia on a visit to Marshall Tito with Randolph Churchill when the aircraft overran the runway. Although a tragic incident, Guy was fortunately the only fatality.

I had already been accepted, subject to passing the

entrance examination, for the Royal Air Force College, Cranwell. However, this method of entry ceased to become an option because of the war, when a massive recruitment drive commenced. The alternatives were to join up like thousands of others or to enter via the University Air Squadron scheme. I chose the former as I thought it would get me into action more quickly. This assumption proved to be incorrect and there was in fact no difference in the time it took.

I shall never regret my choice because, in the space of three weeks, I moved from the sheltered life at Winchester into the real world where I was exposed to all sorts of different people and I benefited from this exposure greatly. This experience has been of enormous value to me throughout my life.

THREE

The Royal Air Force

I joined the RAF at Lord's Cricket Ground on 17 August 1942, just under three weeks after leaving Winchester. Lord's was known as ACRC (Air Crew Receiving Centre) and literally thousands of would-be air crew passed through it. We were met by Flight Sergeant Barnard who addressed the eager young faces with 'My name is Barnard, but you can call me Bastard!' He had a extensive vocabulary of four-letter words and often threatened to drive his '****ing fist' through your head, in an attempt to instil some sense into you, if you unintentionally turned the wrong way about during drill.

We were billeted in blocks of flats around Lord's where we stayed for about two to three weeks until we were fully kitted out with uniforms. This was followed by several months split between an Initial Training Wing at Newquay and a battle course at Ludlow. This period covered lectures in navigation, airmanship, meteorology, signals, Air Force law and mathematics. At the end of which those who were classified as UT (under training) pilots went to a grading school for five hours flying in Tiger Moths. This confirmed one's suitability for further pilot training or alternatively for navigator, bomb-aimer or air-gunner duties.

The majority of flying training during that time took place overseas in Canada, Rhodesia and the USA and we were dispatched through an RAF centre outside Manchester. I was drafted to the USA on the *Queen Mary*, via Monkton in Canada, followed by four days by train to Phoenix, Arizona.

The flying school, No. 4 British Flying Training School, was located at Falcon Field, Mesa about 30 miles out of Phoenix. We started primary flying on Stearman biplanes (PT17A) followed by advance flying on the Harvard (AT6). We were expected to solo in around five hours, otherwise the inevitable elimination flight took place. This resulted in a large 'chop' rate and indeed on my course over 50 per cent were eliminated for various reasons.

My primary instructor, Ray Shelton, was an emotional type of man, who tended to shout and curse over the intercom when foolish mistakes were made. I was coming up to five hours flying without going solo and was told in no uncertain terms that I was not doing too well, so I took the bull by the horns and said to Ray 'How about letting me have a go on my own?' Ray stood up in the cockpit, throwing his parachute and helmet away saying, 'You'll break your neck you ****ing son of a bitch!' And so I took off on my first solo flight and never looked back.

The advanced phase on the Harvard (AT6) started after sixty hours on the Stearman under a delightful man called Norman Poteet with whom I got on extremely well. I took four hours to solo on the Harvard, the aeroplane on which we learnt aerobatics, navigation, formation flying, night flying, stalling, spinning and all general flight manoeuvres, until completion of the course at 200 hours, which included a cross-country flight to Texas. At this point successful candidates were presented with their wings and were commissioned as Pilot Officers or classed as Sergeant Pilots. In my case I was commissioned being fourteenth out of seventy in the Order of Merit. After an excellent party and some sad farewells my course boarded the train back to Monckton to await a ship home. We sailed from New York on 23 December 1943, once again aboard the *Queen Mary*, but this time there were many thousands on board because of extensive US troop build up prior to D-Day.

On arrival in the UK our flying prospects looked rather

gloomy as hundreds of surplus air crew were hanging around the Receiving Centre at Harrogate. This was because Bomber Command losses, although appalling, were not as high as had been anticipated by the Air Force planners. So it was a question of filling in time with more battle courses, flying refresher training on Tiger Moths and any other duty which could be dreamt up.

I was fortunate to find I was posted to a Canadian Air Force OTU, where the Canadians were quite clear from the beginning that if I wanted to fly I should do so. In fact they saw to this in no uncertain way, allowing me to check out on the Wellington. I used to carry out air tests and take part in a number of special operational sorties when 'Screen' (instructor) crews were used to supplement the main Bomber Force.

After five months the RAF caught up with me and I was sent on another refresher course on Tiger Moths. The situation soon improved enabling me to complete an Advanced Flying Course on Oxford twin-engined trainers, before being posted to No. 46 Squadron equipped with Stirlings. I also had a period of glider towing at RAF Netheravon.

As soon as the war in Europe ceased, Stirlings Vs, developed from the glider tug Stirling IV which itself developed from the Stirling III, were used to ferry personnel and supplies from India and the Middle East along with several other aircraft types, mainly Yorks, Dakotas and Halifaxes. No. 46 Squadron was based at Stoney Cross, Hampshire but used RAF Lyneham for staging passengers. I remained with 46 Squadron for eighteen months. When the squadron re-equipped with Dakotas, I remained at Stoney Cross for a time piloting Stirlings from and to Northern Ireland where they were broken up. The Stirling was a lovely aircraft to fly once in the air, with a phenomenal rate of roll for its time. Its basic handling qualities had been ably demonstrated by a half-scale wooden prototype, the S31 known as the 'little bomber'. The original concept to meet the Air Staff specification B12/36 had a wing span of 112 ft in order to provide good high-altitude performance. Unfortunately, before any prototypes were ordered, the Air Ministry specified that the wing span must not exceed 100 ft because the standard RAF hangar could not accommodate anything larger.

By this stipulation the Air Staff wrecked the capability of the Stirling. Flight test results from the S31 produced a recommendation to increase the wing incidence by 3° but production tooling for the Stirling had passed the stage when such a change could be accommodated. A compromise of increasing the ground angle by 3° through lengthening the landing gear was adopted. This resulted in the Stirling having a long and spindle-like landing gear which gave endless trouble including total collapse on the very first flight. In spite of the change, the performance of the Stirling I was found wanting and the altitude capability with a full bomb load was limited to about 13,000 ft. This compared very unfavourably with the Halifax and especially the Lancaster. Consequently, heavy losses were sustained by the Stirling squadrons although the aircraft demonstrated its remarkable ability to survive major battle damage on numerous occasions. The net result was that the Stirlings were gradually phased out of the main bomber force and used for special operations, paratrooping and supply dropping and glider towing. The Stirling IV was built as a glider tug and towed either one Hamilcar or one or possibly two Horsa for assault duties. RAF Fairford, later destined to be the home of Concorde, was a main Stirling base and the squadrons played a major role on D-Day and also at Arnhem, at Nijmegen and in the Rhine crossing.

The Stirling V was developed from the Stirling IV and had a capability of carrying up to forty troops or twenty fully equipped paratroopers, as well as several other combinations of stretcher cases, jeeps or 6-pounder field guns. I became very attached to the Stirling and built up a considerable number of flying hours on the type.

Up to my time in No. 46 Squadron, my flying assessments had been good, average or above average, but on leaving the squadron I was delighted to be given the coveted rating of 'exceptional', a standard which I maintained for the remainder of my years in the RAF.

In 1946 I had a particularly significant year. Firstly, cricket within the RAF was reborn and I found myself selected for the RAF XI against Worcestershire, who beat us rather easily. Their two England bowlers Reg Parks (fast) and Roly Jenkins (leg-breaks) were a bit too much for the likes of myself, who only contributed one run in each innings. However, this defeat was followed by weeks of matches against many clubs, schools and other military establishments, culminating in the inter-service matches against the Navy and the Army at Lord's. The RAF had a number of ex-professional and University Blues available and were able to field a strong side. A group of us under Air Vice-Marshal Bobby Sharp and captained by Alan Shirreff toured the country and Germany with enormous enjoyment.

Secondly, I was ordered to report to the Air Ministry for an interview. This turned out to be a selection process for the King's Flight, which was being re-formed at RAF Benson. The idea was to select eight Transport Command Captains to complete the four crews flying Vickers Viking aircraft. The four senior pilots being the captains and the four juniors being the co-pilots. The interviewing officer turned out to be the Captain of the King's Flight, Air Commodore 'Mouse' Fielden, who had shaken his fist at me on the beach at Pembrey some twelve years before. I was thrilled to learn that I had been successful especially as, due to my cricket commitments, I was allowed to delay my arrival at Benson by a month so that I could play at Lord's. Therefore, I joined the King's Flight on 5 September 1946 to find that I was detailed for a five-hour cross-country flight to Helgoland, a small island in the North Sea off the German coast, at 0500 hrs the following morning.

'Mouse' Fielden had had a most distinguished career in special operations known as 'cloak and dagger' at RAF Tempsford. He was a great character and one to be feared when his moustache twitched at each end. Stories of his experiences are legion and he had many choice expressions: on one occasion in Malta, I was having some trouble with the BOAC catering manager when 'Mouse' Fielden turned to me and said 'The trouble with him is that he is one of the posh buggers who spells F with two small fs.' 'Mouse' did not see eye to eye with the Station Commander at Benson, mainly because of the endless arguments about whether or not the King's Flight Officers should do station duties, because it interfered with our heavy flying programme.

The Flight was equipped with four new Vickers Viking aircraft, two VIP because the King and Queen did not fly in the same aircraft, one thirty-seater and one fitted up as a workshop with hundreds of spare parts. At that time, unlike today, separating the Royal Family into different aircraft was normal practice.

Our first duty was the Royal Tour of South Africa. The Commanding Officer was Wing Commander W.E. 'Bill' Tacon DSO, DFC, AFC of Hudson and Beaufighter fame in the Second World War. Bill came from New Zealand but served in the Royal Air Force for many years until he retired as an Air Commodore. He was a brilliant pilot and carried out all the conversion flying for the other pilots himself. The other captains were Squadron Leader 'Whacker' Payne, Flight Lieutenant 'Tubby' Welch and Flight Lieutenant Jock Bryce with whom I was crewed. The other co-pilots were Flight Lieutenant Alan Lee, Flight Lieutenant Titchbrook and Flight Lieutenant Richmond. Each crew consisted of the Captain, co-pilot, navigator and wireless operator. Bill Tacon demanded the highest possible standards, so that all the pilots would be rated 'A' category by the Transport Command Examining Unit.

During the course of our preparation for the examinations, a very unfortunate event occurred. A group of us, under Jock Bryce, went to the Transport Command Conversion Unit at Dishforth to find out what some of the questions likely to be asked by the Examining Unit would be. We had flown there in a de Havilland Rapide but were urgently recalled as the aircraft was needed for another duty. We needed to refuel the Rapide before returning to Benson but, unfortunately, the only fuel available to us was of a higher octane than was normally used for this particular aircraft and what resulted was a higher fuel consumption than normal. I first became aware that something was wrong when I found someone sitting at my feet as I dozed in the passenger seat and, on asking what was going on, was told we were about to crash which was when I realized there was no sound from outside as both engines had stopped.

Jock Bryce did a wonderful job negotiating a forced landing in the dark, without any landing lights, although, unfortunately, the aircraft was smashed to bits as we hit some trees on the way down. There were seven of us on board and, incredibly, no one was hurt at all. The crash occurred about 7 miles short of Benson and we were rescued by some Army personnel who were escorting German Prisoners of War to their camp in army trucks and they took us to Benson.

When we arrived at Benson we went into the Officers' Mess where we saw Alan Pudsey, who was regarded as the station 'gen' man. Alan asked me what had happened and when I told him he said that he did not like the sound of it and to meet him out the back of the Mess in ten minutes. I joined him at his car, only to find that he was loading petrol tins, ropes and axes: 'We are going to burn it!' he said. So we returned to the crash site, only to find that some rather fierce Alsatians with their corporal handlers had beaten us to pole position. We therefore diverted to the local pub in Dorchester where we settled into a long session of lament.

One of the members of the Flight at that time was seeing rather more of the barmaid than just her ability to pull pints.

Their secret assignations took place in his little MG where, because of the restricted space, the steering wheel had to be taken off and hung on the door handle outside. One night the steering wheel was removed by other members of the Flight as they drove very slowly past his car, which meant he had to drive back to Benson using a spanner, where he found his steering wheel resting on his pillow.

The Rapide incident was blamed on Jock Bryce and as a result he was unable to stay with the Flight as any blemishes were not tolerated by 'Mouse' Fielden in an attempt to avoid any adverse publicity. However, 'Mouse' always used to stand by his team and, appreciating Jock Bryce's great talent, arranged for him to go to Vickers-Armstrongs where Mutt Summers was in need of another test pilot. Jock Bryce was replaced by a very experienced, although rather reserved individual called Harrison and I remained his co-pilot. Despite the fact I had an 'A' Category Rating, 'Mouse' believed I needed more experience before being made a captain in The Flight.

The tour to South Africa was not without incident. The Bristol Hercules engines powering the Vickers Viking surged during take-off at high-altitude airfields during one of the proving flights. A last-minute modification to the carburettors saved the day, otherwise the flying part of the tour would have been cancelled. Most of our navigators were accustomed to more sophisticated navigational aids than were found in South Africa and had some difficulty with dead-reckoning and map-reading methods. Generally the tour was not very arduous and was in fact most enjoyable, Cape Town acting as our main base for about three months. All four aircraft were normally flown simultaneously when the Royal Family moved about. The flights were known well in advance and there was only one occasion when we were nearly caught short, struggling to find a crew to fly HRH Princess Margaret around Table Mountain. Invitations to several of the Royal balls presented new experiences when it came to filling the dance partner cards. One member of the Flight used his card to chalk up the number of drinks that he consumed during the course of the evening. This alternative use of a dance card reached the ears of the Queen who wanted to know if they all did that. We were all highly intrigued when Prince Philip arrived unexpectedly at Princess Elizabeth's (Queen Elizabeth II) twenty-first birthday ball in Cape Town, an outstanding occasion to which all of the air crew were invited.

When we left Cape Town, the Royal Family were given a number of presents including two Rhodesian Ridgeback puppies and a leopard. I was appointed custodian for the flight home but managed to leave the leopard behind. I had endless fun with the puppies, Banshee and Hoolie, particularly in Khartoum where the very hot temperatures convinced me I needed somewhere better than the Transit Mess to babysit my Royal charges. When we left very early in the morning the VIP quarters, which I had demanded and got for the safekeeping of the puppies, looked as though a bomb had gone off. Water everywhere and some sheets and blankets that, thanks to the puppies, were past their best. The puppies behaved impeccably in the aircraft and I was sorry to see them go, especially into quarantine, when we arrived home. Sadly, Banshee died in quarantine, but Hoolie roamed Sandringham for many years.

When the tour of South Africa was over the size of The Flight was cut by one half to two crews, which left just four pilots. 'Mouse' Fielden was determined to sell air transport to the Royal Family and so decided to acquire two helicopters to deliver the mail to Balmoral during the Royal stay there. The two junior pilots, of which I was one, were sent to learn how to fly helicopters at Beaulieu. We learnt to fly the Sikorsky R4, known as the Hoverfly Mk 1, for this purpose. At the end of the conversion course, Alan Lee and myself ferried the two helicopters to Dyce airport, Aberdeen to await the arrival of the Royal Party at Balmoral. With a top speed of 60 mph, the journey from Benson took about thirteen hours with numerous stops.

On the first Sunday Alan Lee and myself were asked to lunch at Balmoral in order to acquaint ourselves with the landing site. After lunch, we were sent back to Aberdeen in order to fly the helicopters to Balmoral so that they could be inspected. This inspection proved quite successful until Prince William of Gloucester put his hand on the exhaust of my helicopter. He let out a bellow which was heard throughout the Highlands and rushed off to be comforted by his mother. I stood by looking thoroughly embarrassed.

The mail was flown from London to Aberdeen by Viking and then transferred to the helicopters. On the first day of operation Alan Lee and myself arrived at Balmoral at about 0700 hrs where we were met by the Court Postmaster, who used to get very upset if you just called him Postmaster. We were taken by him to the senior servants hall where we enjoyed a very hearty breakfast. When we got back to Aberdeen, all hell had broken loose because the Queen was most upset that we had gone to the servants' hall and, in future, we would have breakfast in the dining room. On these occasions one hoped someone would pass the toast and butter, as one was too nervous to ask.

An invitation to the Gillies' Ball came up which tested me considerably. Scottish dances were not my forte, but I found I was trapped in the dancing area until the whistle blew at half time. After which I had a couple of stiff drinks and did not go back for the second half. On our last night Alan Lee and myself were asked to a private dinner party with the King, Queen and Princesses, as well as the whole entourage. After the ladies left the dining room, we sat either side of the King, drinking what seemed like rather a lot of port while the King asked us numerous questions. When we rejoined the ladies we played some interesting party games including the Feather Game, which involves rolling around on the floor, blowing at a feather and trying to keep it airborne. I was feeling somewhat weary towards the end of the game and saw a very inviting chair which I could sink into, until one of the ladies-in-waiting said to me 'For God's sake don't sit there, it's Queen Victoria's chair and it's reigning Monarchs only!'

A number of visitors and Royal staff used to fly to Aberdeen on the mail run, as it was known. They included the King's hairdresser from Trumpers who used to stay overnight at the Officers' Mess at Benson as the flight to Dyce, Aberdeen was a very early start. On one occasion our number one navigator asked if he could have a quick clip on the way to Dyce, to which the hairdresser replied, 'I only cut gentlemen's hair!' He changed his tune somewhat when he finished up in hospital in Aberdeen, following a crash on take-off. The Vikings had given largely trouble-free service but on this occasion, 12 September 1947, VL245 experienced a runaway propeller on take-off from Aberdeen, which necessitated a forced landing in a field, demolishing a stone wall in the process. One of the crew was seriously injured and most of the passengers, including the hairdresser, had cuts and bruises. The aircraft underwent extensive repairs but was never returned to the Flight. The accident came as a great disappointment after the very successful tour of South Africa and Rhodesia, when 160,000 miles were flown without incident.

By July 1948 preparations were in hand for the forthcoming Royal Tour of Australia and New Zealand and two standard RAF Vikings were acquired in place of the damaged aircraft. In the meantime, an Avro York was loaned to the Flight for transporting the Duke of Gloucester to and from Ceylon. The inclusion of a York in the Flight had been conceived when the Flight reformed in May 1946, but had not been followed up. However, the Ceylon exercise was not repeated and the Flight continued to use the Vikings.

The projected tour to Australia and New Zealand required expansion in aircraft and crews. To our surprise there appeared to be six captains for five aircraft, so it looked as though someone was surplus to requirement. This was not good for morale and eventually Bill Tacon extracted from 'Mouse' Fielden that it was Bill himself who would not be having a crew of his own, but would act as Deputy Captain of the Flight and supervise all the crews. Bill was very disappointed by this decision but he was admittedly looking unwell and over-worked. However, none of the plans came to fruition as the tour of Australia and New Zealand was cancelled due to the King's ill-health in November 1948. Once again this resulted in a drastic reduction in the size of the Flight, which reverted to two crews. Those who had been in the Flight from its reformation in 1946 were all posted to various RAF units. In my case I went to the Special Project Squadron of the Empire Flying School at Hullavington, flying mainly Lancasters, Lancastrians, Meteors, Harvards and Oxfords, developing new techniques for flying in all weather conditions.

Hullavington's runways were rather short for Meteors, but we managed to get by most of the time. However, one of the students making his first flight in a Meteor (there were no dual aircraft) from the short runway, about 1,100 yards, was told to make sure that he landed at the beginning of the runway. This indeed he managed to do, but knocked the landing gear off in the process. He then conducted a 'go-around' on the engine nacelles asking the Air Traffic Control tower if his landing gear was down. The 'wag' on duty replied 'Sure it's down, it's down here!'

Another of the squadron's duties was to assess the new Calvert approach lighting system at London (Heathrow) airport using Lancastrians. This involved operating in thick fog, which did not always materialize according to the weather forecast. The London airport Ground Control Approach (GCA) was invaluable in making this exercise possible and we had complete faith in them.

Towards the end of 1949, the Royal Air Force decided to

form the Royal Air Force College out of the Empire Flying School at Hullavington, the Empire Air Navigation School at Shawbury and the Air Armament School at Manby. Manby was the chosen site as it had excellent buildings and accommodation to absorb the new college. However, its runways were, needless to say, short and no better than Hullavington. The first course comprised mainly group captains and wing commanders, several of whom where badly out of flying practice, to the extent that operating Meteors and Vampires from Manby led to numerous minor accidents due to people landing short of the runway. This led to a decision to open a satellite airfield at Strubby nearby in order to have a longer runway available for jet operations.

The Special Projects Squadron at Hullavington was renamed Research and Development Squadron and carried on as before. A new task of creating an instrument rating test applicable to jet aircraft arose using Meteor 7 dual aircraft. Eventually a check visit was made by two instructors from the Central Flying School (CFS) and the rear cockpit of the Meteor was fitted out with blind flying screens, which prevented the 'pilot under test' from seeing out. On one sortie a single-engine landing had to be made, which resulted in the aircraft missing the runway, tearing across the grass and finishing up in one of the aircraft dispersal areas. This was exciting enough but the next sortie proved even more chaotic. The CFS pilot landed short on to a concrete compass swinging base just before the runway threshold. I was in the rear seat, quite happy and oblivious to what was going on when a voice in front said 'I think I've made a balls of this one!' With that the aircraft banged into the concrete, tearing the main landing gear off. After this flight CFS decreed that the jet instrument rating test was satisfactory. Shortly afterwards I was sent to Westland Aircraft Co. Ltd. at Yeovil in order to convert to the Westland Sikorsky S51 (Dragon Fly) helicopter. The RAF had decided to send two helicopters to

Malaya in 1949 to counter the rebels in the jungle. When it came to finding RAF pilots with helicopter experience, it transpired very quickly that there were none, apart from Alan Lee and myself, resulting from our experience in the King's Flight.

My heart fell a mile at this development because I had remained in constant touch with 'Mouse' Fielden since leaving the Flight and he had already started pointing me in the direction of test flying. De Havilland had no vacancies at the time of his enquiry but Vickers-Armstrongs (Aircraft) Ltd. did have one. This suited me very well because the use of the Vikings in the King's Flight had given me the opportunity to get to know Mutt Summers, the Chief Test Pilot and George Edwards, the Chief Designer and others quite well. Mutt Summers offered me a job, but my application to leave the RAF was rejected because of the forthcoming helicopter operation in Malaya. Fortunately for me, Mutt knew the Air Member for Personnel, Sir Leslie Hollinghurst, very well and between them they fixed my retirement to the Air Force Reserve.

This all took time and a certain amount of confusion arose at Vickers regarding my release, to the extent that on the very day my release came through the RAF system, I got a letter from Vickers withdrawing their offer. Mutt Summers was overseas, so I did not know what to do. In the end I telephoned George Edwards and presented my predicament. 'Leave it to me old son!' he said, which I did and all was well. My enormous debt to George Edwards started on that particular day. George was not only my boss in various capacities but he also became, and is, a real friend. Someone to whom you can turn to for almost any sort of advice. His staff had the greatest admiration for him and he was recognized internationally as one of the real 'greats'. He did not suffer fools gladly and he had some wonderful expressions to cap his feelings. I remember someone pontificating for some time until George's patience was exhausted, so he turned and said, 'And I will tell you something else, if your auntie had *****, she would be your uncle'. On other occasions he would say 'I know its raining, I want someone to tell me how to stop it!' George Edwards showed a tremendous understanding of test pilots and was proud to admit it. We were very lucky to have such a person at the top.

FOUR

Arrival at Vickers-Armstrongs (Aircraft) Ltd.

Iofficially joined Vickers-Armstrongs (Aircraft) on 1 May 1950 when I reported to the Commercial Director, Jack Anderson, armed with the letter confirming my appointment as experimental test pilot. It was a big day for me and I felt thoroughly elated at the prospect of beginning a new career as a civilian. With a salary of £1,200 per annum I thought that I was enormously rich compared to my RAF salary of about £900 per annum. This myth did not last for too long before I had to ask my aunt, Dame Gwendoline Trubshaw, to guarantee an overdraft of £100 at the bank.

When I walked into Jack Anderson's office, I was somewhat surprised to find that his trousers were under repair. His secretary, kneeling on the floor in front of him, was stitching a fly button on – *in situ*. He was acutely embarrassed to start with but, the three of us, soon managed to laugh it off. Jack Anderson was a rather large, jovial individual and I grew to like him very much indeed. Once my meeting with Jack was over, I went to Wisley airfield, just south of Weybridge on the A3 Portsmouth road and opposite the famous Wisley Horticultural Gardens, where I met Mutt Summers.

All flight test operations were carried out from Wisley under the direction of Captain M.J. (Mutt) Summers. Mutt was a very famous test pilot and during the twenty-five years he had been Chief Test Pilot he had known over forty prototypes. At this time prototypes were built very frequently and quickly in response to the many new designs that were produced. Fortunately, I got on extremely well with Mutt, although all the test pilots regarded him with some awe. He was a great pilot, who belonged to 'the seat of the pants' vintage, a great character, full of humour, as well as being an excellent teacher who was always ready to pass on his vast experience.

Jock Bryce, whom I had flown with in the King's Flight, was also there. The Chief Production Test Pilot was George Lowdell and one other pilot, Brian Powell, who had been taken on by the General Manager, B.W.A. Dickson, when Mutt was away and it looked like I was not going to be released by the RAF. There was not much love lost between Mutt and B.W.A. Dickson. George Lowdell was a wonderful character, who consumed gin and French vermouth any time there was a party. He had a splendid bulldog called Caesar, whose idea of heaven was mud and water, which I found out to my cost when I had to share the back seat of George's car with him once. I was caked in it, from head to foot. Brian Powell was understandably not over-pleased when I arrived because my appointment relegated him to production flying.

Our office block was a cottage, which housed Mutt and his secretary downstairs and Air Traffic Control, which included one of the first female air traffic controllers, and two offices upstairs. An extension had also been built at the back of the cottage which accommodated three or four technical staff.

The runway at Wisley was around 6,600 ft long and grass as opposed to tarmac. The airfield was looked after with loving care by the groundsman, Harry Gray, who I soon found was the key player when it came to deciding whether or not flying could or could not start after heavy periods of rain.

When I first arrived, the aircraft at Wisley were the Viscount 630 prototype powered by four Rolls-Royce Dart engines, a similar airframe powered by two Rolls-Royce Tay engines (developed from the Rolls-Royce Nene), a Viking with two Rolls-Royce Nene engines, two Varsity twin-engined trainer prototypes and a flow of production Valetta military transports coming off the production line at Weybridge. My first few months at Vickers were spent getting to know the aircraft under the watchful eye of Mutt Summers.

The Nene Viking did not fly very often, because the aircraft's conventional landing gear – main and tailwheel – supported jet pipes that were sufficiently close to the ground to set fire to the grass surface of Wisley, which did not go down too well with the groundsman Harry Gray. However, the Nene Viking did make the record books when it reduced the flying time from London (Northeast) to Paris (Villacoblay) to one hour.

In June 1950, I was dispatched urgently on an overseas assignment. An Attacker naval fighter built at Vickers-Armstrongs Supermarine was on a demonstration tour of the Middle East, flown by Mike Lithgow, with a Valetta transport in support. The pilot of the Valetta, Archie Boyd, was a member of the sales department at Weybridge and had fallen sick. Jeffrey Quill, who was managing the tour, had retired from test flying, but had flown the Valetta from Basra to Baghdad without being approved by the Ministry of Supply. This caused some stir at Headquarters and I was sent to operate the Valetta for the rest of the tour. In fact I only covered Baghdad to Damascus and Damascus to Ankara, where a smiling Archie Boyd, recovered from his illness, met us on the tarmac. One more leg to Athens was enough to convince me that I was surplus to requirements, so I returned home. The use of pilots in the sales department ceased shortly after this trip as Mutt Summers and those concerned could not agree that using pilots from the sales department on sales trips was beneficial. Mutt felt that he did not have the same amount of control over them as he had with the test pilots.

I returned from my first overseas trip to play in two annual cricket matches, one against the Contracts Branch of the Ministry of Supply and the other against the Operational Requirements of the Air Ministry (now Ministry of Defence). Our side was captained by our Chief Designer, George Edwards, who was a fine cricketer himself, and consisted of members of the Director's Executive Mess. In the match against the Contracts Branch, it was unfortunate that I had not been briefed that the Director of Contracts had to be allowed to make 50 runs. He opened the innings and I had him caught at the wicket, first ball, by Brian Powell. This went down very badly with our Managing Director, Sir Hew Kilner, who remarked during the after match dinner that he detected an unnecessary seriousness entering the contest. Fortunately, George Edwards once again came to my rescue and I did not get the sack. These matches, especially against the Operational Requirements Branch, continued for many years until the advent of the M25 motorway significantly reduced the cricket ground on the Vickers Sports field. Up to that time the ground was up to County standard and was used by Surrey County Club at least once per season.

These encounters on the cricket field were just a part of a life that was very amiable. I was happy to be at Vickers-Armstrongs and I liked most of the people with whom I worked and came into contact with enormously. I had no regrets about having retired from the RAF at my own request, although some of my closest friends had said I was a fool to have done so as they believed my prospects in the RAF would reach much higher levels that I would ever attain in the aircraft industry. I still feel that I made the right decision.

FIVE

The Vickers-Armstrongs Viscount

When I first joined Vickers-Armstrongs my main work revolved around the Varsity prototypes, a twin-engined trainer built for the RAF, until they went to Boscombe Down for their service acceptance trials and the Tay Viscount. This was an experimental aircraft which had been built from the second type 630 fuselage. It was supposed to be used for research flying. In fact its main use was as a test vehicle for the hydraulic operated powered flight control system for the Valiant four-engined bomber. More was probably learnt from the actual installation of the system and its problems than from any actual flying. The Tay Viscount helped to keep the Viscount in the minds of the Government and British European Airways (BEA).

In 1950, I flew the Tay Viscount at the Farnborough Air Show. I led what was known as the heavy circus – this was not very demanding, consisting of take-off, flypast and landing. It was, however, my first major air show and I was quite excited as we taxied out in front of a sizeable crowd. The other aircraft were engine test bed variations and other prototypes that had appeared in earlier shows. The Viscount started life as the VC2 subsequently named Viceroy until August 1947 when it was renamed the Viscount – 'an alteration not entirely unconnected with political events at that time', which were associated with Indian independence.

The Brabazon Committee had been formed under Lord Brabazon to examine future civil aircraft projects and Vickers had already made a firm proposal for an aircraft to the

Brabazon Committee Specification 11b. The proposal centred around the use of the new turbo-prop engine, the Dart being developed by Rolls-Royce.

In May 1946, the Ministry of Supply ordered two prototypes specifying the Armstrong-Siddeley Mamba engine to carry thirty-two passengers. The same engine was chosen to power the Armstrong-Whitworth Apollo. Vickers was prepared to build a third prototype at its own expense using the Rolls-Royce Dart but this never materialized as the official decision for the Mamba was changed to the Dart a year later. In retrospect the choice of the Dart made the Viscount into the fine aircraft that it became while the Apollo soon ceased to be a serious competitor.

The Dart flew in a number of test bed aircraft, a Dart-Wellington and a Lancaster. All seemed to be going very well with a very close working relationship between Vickers and BEA to produce an aircraft with increased passenger comfort and additional safety due to the use of kerosene instead of petrol. There were, as is so often the case, some who doubted the economics of the Viscount with its new turbo-prop engine designed to cruise at about 25,000 ft, thus needing a pressurized fuselage.

Lurking in the wings during this period was a twin-piston engine aircraft, the Airspeed Ambassador and its manufacturer was pressuring BEA for an order. The order came at the end of 1947 for twenty aircraft, making prospects for the Viscount look somewhat gloomy, but the Ministry of Supply were still interested in it as a military transport. Also, in BEA there was always strong support for the Viscount from Peter Masefield (now Sir Peter Masefield), Chief Executive. The support for the Viscount within the Ministry came from Sir Alex Coryton, Controller Supplies (Air) and Sir Cyril Musgrave. Inside Vickers, Sir Hew Kilner was another supporter but it was the invincible faith of George Edwards which kept the project alive.

The first flight of the prototype 630 took place on 16 July

1948 with Captain Mutt Summers at the controls, marking the first flight of an aircraft powered by four turbo-prop engines. After some fifteen hours test flying, the aircraft eventually appeared at the SBAC Air Display. The aircraft carried out extensive flight trials covering handling, tropical trials and severe icing tests. The smoothness of the aircraft in cruise at 25,000 ft was considered remarkable. This was followed by a series of proving flights between London (Northolt) and Paris flown by BEA for a two-month period and an extensive demonstration tour made to all the European capitals in March and April 1950.

Design studies continued and largely due to increased power becoming available from the Dart, George Edwards decided to increase the size of the fuselage by 6 ft 8 ins and wing span by 5 ft, which resulted in the 700 Series. Vickers prototypes used to be built at Fox Warren, a hangar located in the woods between Weybridge and the test flying base at Wisley. Once assembled the aircraft had to be towed down the A3 to Wisley for final assembly and preparation for flight, which was quite a performance. The trips had to be made at about 0500 hrs on Sunday mornings to ensure not only that the road traffic was at a minimum, but also the telegraph poles had to be dug up to make way for these unusually wide loads.

The first flight of the 700 Series was also made by Captain Mutt Summers, with Jock Bryce as co-pilot, on 28 August 1950 from the main factory at Brooklands to Wisley. I managed to persuade Mutt to let me squeeze behind his seat. For me this was an epic occasion as I had never witnessed a first flight before and I was determined to learn all that I could. First flights always seemed to coincide with opening time at the Wisley Hut Hotel where a gathering would take place in order to celebrate a successful flight with some gusto and with some very sore heads the following morning.

Later in August there was great rejoicing when BEA gave a production order for the 700 Series, together with BOAC for their subsidiary British West Indian Airlines. In September there was further success when the 630 prototype was given a Certificate of Airworthiness although it did not cover pressurization or flight in icing conditions, but both these followed shortly afterwards.

Flight development of the 700 Series lasted for approximately two years, with most of the flying done by Jock Bryce and myself. The programme covered full handling and performance tests, tropical trials, cold weather trials and de-icing trials, as well as numerous demonstration flights and tours. Bearing in mind that the certification of the 700 was a major step forward, the tests went very well.

Meeting the requirements for stalling characteristics necessitated the fitting of a stick shaker to give adequate warning before the stall. Stalls in turning flight were a bit exciting on occasions and indeed an inadvertent spin resulted from one of them, when the aircraft was being flown by D.P. Davies of the ARB. Turning flight stalls on the later 800 Series were even more difficult. Rate of roll with one engine stopped at take-off safety speed was very marginal and in later years was regarded as being the minimum acceptable – about 7 degrees per second as I recall.

An unpleasant and very difficult technique was used for measuring landing distance. The aircraft was positioned at about 1,000 ft on the approach, with landing gear and landing flap down. Power was then selected to idle and the remainder of the approach flown in this manner, normally at 1.3 x the stalling speed. A high rate of descent was inevitable with a high degree of skill and judgement necessary to execute a touch down that did not drive the landing gear up through the wings. The approach being between 7° and 8°.

The 700 prototype was damaged during one of these tests while an 800 series a few years later tried the same thing in Johannesburg during tropical trials and returned to the UK on a boat. When I was detailed to conduct the tropical trials on the Vanguard in December 1959, I refused to adopt this technique, which was supposed to give the shortest landing distance and insisted on using a normal approach, of about 3° to 4°. Gradually sense prevailed and the Air Registration Board agreed to what became known as the Rational Landing Technique.

The era of the Viscount saw the arrival of the Air Registration Board's new chief test pilot, D.P. Davies, an ex-naval test pilot with whom I had served at Hullavington. Dave Davies had a considerable impact on civil aircraft certification for many years to come. Dave's word was seldom questioned by manufacturers during certification but he tried to be fair in handing out the same treatment to everyone.

Production orders for the 700 followed that from BEA, namely Air France (March 1951), Aer Lingus (November 1951), Trans Australia Airlines (May 1952), Trans Canada Airlines (November 1952), BWIA (1952), and many others worldwide. The certification date of the type 700 took place in December 1952 with the first delivery to BEA in January 1953. In order to build up experience on the Dart engine, BEA equipped two Dakota DC3s with two Darts.

Much of the credit for the Trans Canadian Airways (TCA), then Air Canada, orders goes to George Edwards, who personally convinced the President of TCA, Gordon MacGregor, of the Viscount's virtues in the face of extreme pressure from American manufacturers like Douglas, Convair and so on.

A enormous number of design changes were required to meet TCA's specification and more design man hours were spent on this model than went into the original prototype. However, it resulted in an aircraft suitable for the North American continent, which was a big step forward. Orders continued to pour in, not least from Capital Airlines based in Washington DC who bought sixty. At one time Viscount production rate between Weybridge and later Hurn reached seven aircraft per month.

Further development continued and BEA ordered the 800 Series in February 1952. The main problem encountered on this model was stalling in turning flight to the right when the aircraft either rolled under to a high bank angle of 70° to 80° or flicked out over the top. No final solution was forthcoming in spite of using various wing fences, vortex generators and spoilers but BEA wanted the aircraft. The Chairman of BEA Sir Anthony Milward called a meeting of the Air Registration Board, Vickers represented by George Edwards, Jock Bryce and myself, as well as BEA engineers and training captains on Christmas Eve. He opened the meeting at 1400 hrs by saying it was Christmas Eve and the meeting would close at 1530 hrs. After some discussion, it was agreed that the aircraft could be certified and accepted.

A further version, the 810 Series, came next and was ordered by a number of airlines including Continental Airlines of Denver, Colorado. I have always thought that the 810 Series was the best of all the Viscounts, because it was very solid with a feeling of considerable power. A total of 444 Viscounts were sold worldwide. All these models were powered by the Rolls-Royce Dart engine, which was developed considerably throughout its life. The Dart was and is an outstanding engine and really made the Viscount into the world beater that it became.

The total of 444 new Viscounts were produced by Vickers at Weybridge and Hurn for some sixty operators. The number of operators increased considerably when second-hand aircraft became available. A sale of Viscounts to the Queen's Flight in order to replace the Vikings was considered but did not materialize and the HS748 was chosen instead. The list of operators speaks for itself and makes the Viscount one of the outstanding achievements in the history of British aviation. When production rates peaked at seven to eight aircraft per month between Weybridge and Hurn, Weybridge was producing four Valiant bombers per month as well.

With such a large number of aircraft operating on a worldwide basis, it is inevitable that there have been accidents. Over the last forty-five years over 25 per cent of the 444 Viscounts have been involved in accidents due to a variety of causes. These have included crew training, especially during engine failure on take-off, flap failures, incorrect altimeter settings, while many have been unknown. One particular accident has always stuck in my mind because it shows the fallibility of the human being: the loss of one of Hunting-Clan's V732 Series which took off from Heathrow following overhaul. The Viscount's elevator was equipped with a spring-tab in order to help the pilot move the elevator without having to exert excessive force. The design of this tab and its attachment followed established design principles to prevent incorrect fitment. In this case the mechanic disassembled the main attachment point on the basic structure, thus enabling him to fit the tab the wrong way so that its function was reversed. Soon after take-off, the pilot was confronted by a need to apply an excessive pull force to keep the nose from dropping. The pilot managed to control the aircraft for nearly fifteen minutes before the control forces overcame him and the aircraft crashed near Camberley, killing all six occupants. On certain occasions it was necessary to visit crash sites, which was always a gruesome experience, and I admire the professional investigators who have to do it all the time.

SIX

The Vickers-Armstrongs Valiant

M y release from the RAF was approved on the basis of my intended involvement with the four-engined Valiant V Bomber programme. Vickers had a long history of building large military bomber/transport aircraft and during the inter-war years the capability of the RAF was based very heavily on the Vickers' 'stable'.

High-altitude technology stemmed from the long-range Wellesley, the Wellington VI and the Wellington X; the Wellington itself was Britain's most prolific two-engined bomber during the Second World War. George Edwards was Experimental Manager at Vickers during the war years and was closely involved with the development of design and manufacturing techniques that proved to be of great value in the near parallel development of the Viscount and the Valiant. He was also a member of a team that examined basic theoretical research on swept wings obtained from German sources after the Second World War. Towards the end of the war Bomber Command was due to be completely re-equipped with the piston-engined Lincoln (Lancaster V) and the Vickers Windsor, a four-engined development of the Wellington, but the Windsor never came to anything with only three prototypes flown. One of these was still parked at Wisley when I made a visit there in 1946.

The dropping of the atom bomb on Japan in 1945 ended the Second World War abruptly, but nevertheless influenced military thinking a great deal with regard to bomber strategy after the war. Definitive British jet bomber design studies under Air Staff Requirements OR229 became specification B35/4 issued on 9 January 1947. The specification called for a 3,350 nautical mile range, a maximum operating speed of 500 k (575 mph), at a Mach no. of 0.875 with an 'over the target' height of 50,000 ft. The aircraft had to be capable of carrying a nuclear weapon of 10,000 lbs or a range of conventional bombs internally. The crew were to consist of two pilots, two navigators and one radio/radar counter-measures operator, situated in a nose pressure cabin. Navigation and bomb aiming were to rely on an advanced version of H2S radar. Industry response produced some revolutionary designs which caused some concern among Government technical departments as to their likely success.

Consequently, a much lower specification B14/46 was issued in August 1947 as a safeguard against the possible failure of the B35/46 specification. A conventional aircraft with a straight wing resulted in an award being made to Short Bros to build the SA4 Sperrin. In fact the Valiant flew before the Sperrin, a prototype of which remained at RAF Farnborough for a number of years, and I had a flight in this aircraft in 1955 for general experience. The Vickers response to B35/46 was not quite as advanced as those from AVRoe (Vulcan) and Handley Page (Victor) and was initially rejected by the Air Staff. However, the Berlin Blockade in 1948 jerked the Air Staff into high gear to re-equip Bomber Command as quickly as possible in order to counter the new threat which Russia now posed. The Air Ministry returned to the Vickers proposal and asked George Edwards if he could produce such an aircraft in a much shorter time scale. Dates quoted were for the first flight during mid-1951, the first production aircraft flight by the end of 1953 and the start of deliveries to the RAF by the beginning of 1955. Performance and speed were tabled as being very close to specification B35/46.

The Air Staff were sufficiently impressed by the proposals of George Edwards with his team of Sam Richards

(Aerodynamics), Basil Stephenson (Structures) and Henry Gardner (Stress) that they issued a new specification B9/48 based on the Vickers Type 660 design. The company was instructed on 15 April 1948 to proceed with the construction of two prototypes, for which a contract was placed on 2 February 1949. Many aviation critics have often wondered how the UK could afford the development of three V Bombers, but there was no convincing answer. The Vulcan programme was flown by a great friend of mine, Roly Falk, who had a tremendously distinguished test flying career and who would have succeeded Mutt Summers but for an unfortunate accident in a Wellington fitted with reverse pitch propellers. The Victor test pilots 'Hazel' Haselden and John Allam were also personal friends.

The first prototype was to be powered by four Rolls-Royce Avon RA3 turbo jets and the second by the Armstrong-Siddeley Sapphire. Progress on the first prototype was very rapid and was taking visible shape at Vickers Experimental Shop, Fox Warren when I joined Vickers in 1950. Later in the year the aircraft was transferred by road to the Vickers flight test airfield at Wisley, with final assembly accomplished in less than six months, demonstrating what could be achieved by a devoted team covering many disciplines. Those connected with today's prototypes could learn a great deal from the Valiant story.

The Valiant exhibited a clean shoulder wing layout with a circular fuselage to accommodate the nose-mounted H2S radar and the pressurized five-crew nose capsule. The large bomb bay lay behind the nose-wheel. The wing sweep had a mean 20°, but was more sharply swept inboard than on the main outer wing. While the amount of sweep on the outer section was determined by prevention of tip stalling, the greater amount on the thicker inner section (where the four Avon engines were buried in the wing root area) was required to restore the effects of the sweep lost at the wing body intersection and hence the overall balance of the wing from root to tip. The aerodynamic concept was invented by Sam Richards and patented. The high set tailplane was arranged to be well clear of the jet exhaust efflux. The need to achieve good airfield performance conformed to the general layout and contrasted sharply with American designs of this era (for example, B-52) with their long, thin, high aspect ratio wings with engines mounted externally in pods underneath, where operation from long runways was acceptable.

The Valiant had a bicycle undercarriage which retracted outwards into the wing. It was essentially an all-electric aircraft: landing gear, flaps and movable tailplane were all electrically actuated. The flight controls were operated by Boulton and Paul electro-hydraulic power units with artificial feel provided in order to simulate the normal feel supplied from aerodynamic flight controls. The engine intakes on the first prototype were a letter-box type and were substantially changed on the second prototype when more powerful Avon RA7 engines became available. The Armstrong-Siddeley Sapphire version never saw the light of day, although some ground tests on the engine intake rig at Weybridge did take place.

Two things stand out in my mind before the Valiant's exciting first flight on 18 May 1951. Wisley, as a grass airfield, became completely water-logged at least once a year and on one such occasion Mutt Summers, Jock Bryce and myself were talking to George Edwards in front of the gleaming monster (the aircraft was polished not painted). Mutt said to George Edwards, 'You may still have to take it to Boscombe Down.' I thought George Edwards was going to explode, as the aircraft would have had to have been transported by road, when he replied, 'If I had thought that I would have built it differently!' Not a popular observation from Mutt, and fortunately the airfield dried out in time, but I saw it as the first demise of Mutt. The second memory concerns the aircraft itself. The Valiant's flying controls were power assisted with artificial feel, each power unit being electro-hydraulic and not conventional aerodynamically balanced controls which had always been the case on the many previous prototypes flown by Mutt Summers. The system did incorporate what is called manual reversion with which, if all power failed, the aircraft could still be flown, albeit with degraded flying qualities. So this was something new for Mutt, who horrified us by saying that he might make the first flight in manual so that he could have a better feel. This idea did not last for long but it illustrated that technical developments were at long last outstripping Mutt.

The first flight of WB210 took place from the grass airfield at Wisley on 18 May 1951 flown by Mutt Summers accompanied by Jock Bryce. Like most of Mutt's first flights, it was of short duration with the landing gear down and at low level. Some years earlier Mutt had an accident in a Warwick due to a flying control problem but he had managed to side-slip the aircraft into the trees on the north side of Wisley. In doing so he fared a lot better than some of his fellow test pilots and consequently developed a theory that remaining close to the trees was a wise precaution. It looked fairly horrible from the ground, seeing this great monster making its first flight at less than 500 ft. The flight raised one huge problem: namely that the grass airfield and the Valiant did not go together. The bicycle landing gear had produced ruts deep enough to lie in, especially when the aircraft turned around.

It was immediately clear that an alternative airfield had to be found and, as luck would have it, British Overseas Airways were moving out of their maintenance base at Hurn, Bournemouth which had some excellent hangars. A decision was reached by the Vickers management that we would all move out of Wisley while a tarmac runway was laid. In early June we moved all the development aircraft to Hurn while production aircraft from Weybridge made their first flight to Hurn instead of Wisley and so started the Vickers operation from Hurn which lasted for the next thirty-five years or so. The services of a Valetta transport aircraft were acquired in order to provide a shuttle service between Weybridge, Wisley and Hurn.

The second flight of the Valiant was made by Mutt Summers to Hurn. Mutt went on to complete one more flight on the Valiant and then turned the programme over to Jock Bryce and myself, our first flight taking place in WB210 on 20 June 1951. On 24 June 1951 I flew as Captain for the first time, an occasion which I look back on with much pride as this started my long and close association with this fine aircraft, which occupied so much of my life in later years.

Valiant development proceeded without too many unexpected problems for the next eight months until disaster struck on 12 January 1952. Fire in the starboard wing started by fuel from unsuccessful engine relights in the air entering the wing flap shroud resulted in an initial fire burning through a main fuel line coupling. An intense fire then broke out and severed the aileron control rod in the right wing. It was only when an uncontrolled roll of the aircraft occurred that the crew realized that there was major trouble. In fact the aircraft had been on fire for some minutes before it rolled and now fire was coming out of the wing leading edge. Jock Bryce ordered the crew to bale out before leaving the aircraft himself. Both pilots had ejector seats and the three occupants in the rear of the crew cabin, Roy Holland, John Prothero-Thomas and Jan Montgomery all managed to escape through the main entrance door. The co-pilot, our RAF Bomber Command Liaison Officer, Squadron Leader Brian Foster ejected but tragically hit the fin. Jock Bryce was not seriously injured following his ejection but Brian Foster, although still alive when found, died soon afterwards. The aircraft itself crashed at Bransgore near Hurn. Any accident is tragic and doubly so in this case because a similar fire incident had occurred on a Comet, but it would appear that all the information relating to the incident had never been circulated.

Fortunately, the second prototype WB215, Vickers Type 667 powered by increased thrust Rolls-Royce Avon RA7 instead of Sapphires, was nearing completion at Wisley and actually flew on 11 April 1952 in the hands of Jock Bryce and myself as co-pilot. The combination of the urgency of the programme and the faith in Vickers led the Government to place an initial production order for twenty-five aircraft on 20 April. Therefore, the loss of the first prototype was not such a set-back to the programme as it might have been.

While the Valiant programme continued, the company was also heavily involved in the worldwide Viscount programme. We had been joined by additional pilots Stuart Sloan and Philip 'Spud' Murphy, as well as Guy Morgan who moved across from Supermarine. Viscount commitments took up more and more of Jock Bryce's time, resulting in most flying on WB215 falling on me. This suited me very well as, after all, the Valiant was the reason the RAF allowed me to be released to Vickers.

Brian Foster had been replaced by Squadron Leader (later to become Wing Commander) Rupert Oakley, a heavily decorated veteran of the Second World War, who was a delight to be with and to fly with. The rest of the crew were normally those who had escaped from the first prototype. Understandably each flight was not very easy for them and consequentially nerves were tense on some occasions. During some flights, peculiar noises were reported from the bomb bay or elsewhere, necessitating an early return to base. The crew's determination won in the end and what appeared as a problem disappeared.

On the first climb to 40,000 ft, it became clear that something was wrong with the engine control as it became necessary to keep opening the throttles to maintain engine speed. At the top of the climb it seemed prudent to throttle back to two engines at a time. Just as well, because the first two immediately went out. A return to Wisley on two engines followed, but some intake buffeting was evident.

Following some flights with tufts fitted on the intake lip, it was clear that some fundamental change was required to the spectacle intakes. This was achieved by applying lumps of putty suitably shaped and then covered with fabric. This rather rudimentary technique produced the solution and modified special intakes were made out of wood for fitment to the aircraft. This worked out so well that the wooden spectacles were never removed from WB215 and remained on it throughout the remainder of the flight test programme.

The flying characteristics of the Valiant were very good. However, control forces when flying in manual reversion were very high but acceptable for emergency use. The aircraft had a tendency to Dutch roll at high altitude but this could be stopped by extending the speed brakes and was normally contained by a yaw damper. Flight at high Mach number above M=0.86 produced very heavy buffet, which became intense as the design limit of M=0.92 was reached. In order to demonstrate M=0.92 a steep dive was made from 50,000 ft in order to reach M= 0.92 at about 45,000 ft when the airspeed was comparatively low and this kept the buffet to tolerable limits. At lower altitudes the level of buffet was quite frightening.

During the assessment of the aircraft by the Aeroplane and Armament Experimental Establishment (A & AEE) Boscombe Down, one of their pilots did not follow my recommendation and started the dive from a lower altitude. He was of course quite entitled to do this as he was well within the design envelope. However, the after flight inspection revealed a split in the wing leading edge. Heavy buffet is a hazard in my experience which always ends up with something damaged. The same scenario manifested itself again on the first production aircraft during high-speed tests. In this case the aileron rods broke leaving test pilot, Bill Aston, with no roll control. Due to exceptional skill, Bill managed to land the aircraft at Boscombe Down using rudder and elevator only. Examination of the flight records, which were more sensitive than on the prototype, showed that violent aileron buzz was the cause of the problem. Use of vortex generators fitted to the wing and much stronger aileron rods produced a solution, although the modified aileron rods made the aircraft even heavier than before when flying in manual reversion.

Another critical test was demonstrating the carriage and dropping of the 10,000 lb nuclear weapon, known as the Blue Danube. This was a Top Secret operation and was conducted from Farnborough. The weapon was loaded in a specially screened area, bomb doors closed and the aircraft towed out for flight. Wind tunnel and initial tests indicated that the bomb 'wanted to fly' inside the bomb bay after release. This necessitated the fitment of dragon's teeth in front of the bomb bay which were fixed on the prototype but became retractable on production aircraft. The bomb was carried on a Vickers bomb slip of mature design. After take-off from Farnborough, the aircraft routed south and then east past Dorking in Surrey before turning in a northerly direction to the North Sea ranges off Orfordness for the actual drop. Whereas the release from the aircraft proved to be satisfactory, I noted a definite pause between pressing the release button and the actual release, which caused the aircraft to lurch as the load went. A special recording gauge was fitted to the bomb slip in order to show when the slip released.

On 30 July 1952 I took off from Farnborough for a routine flight but as we passed Dorking I felt the customary lurch. Examination of the bomb slip gauge by one of the Flight Test Observers indicated that the slip had fired. In which case, the bomb had to be resting on the bomb doors. I decided to position over the Thames Estuary and open the bomb doors. On doing so there was considerable graunching and grinding. The bomb had indeed been sitting on the bomb doors until it was able to fight its way out as the doors opened. As I was uncertain of the extent of the damage, I made an emergency landing at Manston in Kent. The damage to the bomb doors was considerable and provoked an American Air Force sergeant to say, 'Those bomb doors look kinda buckled!' I replied, 'You'd look kinda buckled if you'd had the same object sitting on you!' The aircraft type foxed him and I heard him announce to his assembled group, 'I guess it must be a Boeing!' After removal of the damaged doors we flew back to Wisley later in the day.

Another mishap arose in October when trouble occurred on the left landing gear where the electrically driven actuator suffered a clutch failure, which meant that the left landing gear was not locked down. When I landed at Farnborough I knew a landing gear collapse was inevitable, but I managed to keep the left wing up until at low forward speed. A majestic fall of the wing on to the runway caused damage to the flaps but not much else and repairs were made very quickly.

American interest in the Valiant reached a high level resulting in the Chief of Staff United States Air Force, General Hoyt Vandenberg, General Johnson and the famous General Curtis Le May of the Strategic Air Command, accompanied by Marshal of the RAF Sir John Slessor Chief of the Air Staff and Air Chief Marshal Sir Ralph Cochrane paying a visit to Vickers. There never was any real hope that the Americans would buy the aircraft, although I believe George Edwards was put under some pressure to build the Boeing B-47 under licence. Eventually all that resulted was that Le May insisted on side-by-side seating for the two pilots in the Boeing B-52 instead of the tandem arrangement in the B-47 and the B-52 prototype.

The next major step in Valiant development was the fitment of two underwing 'drop' tanks to provide additional fuel. These tanks had a capacity of 1,500 gallons each and the thought of dropping them caused a lot of concern to a number of people including myself. Common sense soon prevailed among the Air Staff and they were fitted as permanent non-droppable fixtures. This major change of configuration necessitated a full exploration of the flight envelope and the increase in maximum take-off weight meant that the Wisley runway was not long enough. So, I found myself spending more and more time operating from Boscombe Down. This had some advantages by creating a very close relationship with the A & AEE Test Pilots (all RAF officers) of 'B' Squadron who carried out service trials on all large aircraft. 'A' Squadron looked after fighters, while 'C' Squadron was naval aircraft and 'D' squadron helicopters.

During the time the underwing tanks were being installed, the question of Bomber Command flying the prototype in the New Zealand Air Race arose. The late Group Captain Hamish Mahaddie gave George Edwards some worries when he announced at an early meeting 'You must realize that when we get started, it will be shit or bust'. As things turned out the aircraft was not ready in time and a Canberra was used instead. In this connection I fell foul of the Controller Aircraft, Air Chief Marshal, Sir John Baker when, following a dinner at Boscombe Down, I suggested it would be better to use a Boscombe Down crew instead of one from Bomber Command. I never had liked the look of him and this episode confirmed my opinion, as this was our first and last conversation.

The next Valiant to fly was the Mark II prototype WJ954, which made its first flight on 4 September 1953 with Jock Bryce and myself as pilots. This was the last aircraft to be built at Fox Warren and taken down the London– Portsmouth road to Wisley for the first flight. There was just time to complete the magical minimum of ten hours for qualification at the SBAC Farnborough Air Show. The Mark II aircraft was slightly bigger than the standard airframe in order to house extra equipment. The bomb bay was a few inches longer and the retracting of the landing gear differed considerably. A new four-wheel bogie retracted rearwards into underslung pods on the wing trailing edges instead of sideways into the main wing structure on the Mark I.

The role was as a 'Pathfinder' incorporating a high-speed capability at very low altitude. In order to fulfil this, the design diving speed was 580 K Indicated Air Speed which meant that the main load-bearing structure had to be beefed up considerably. The artificial feel on the ailerons was changed to spring feel in order to reduce aileron heaviness at high speed. It was painted in black finish hence its nickname 'Black Bomber', the colour scheme arising from a remark that I made at a design office meeting, when I said '. . . and paint the f****r black!'. Other than the high rate of roll possible, it flew like any other Valiant.

The first task was to clear the aircraft for its increased airspeed limits by measuring in flight loads and by showing the structure was flutter (tail wagging dog) free. Smooth air at low level was needed to carry out the latter and flights commencing at 0400 hrs at 1,000 ft over the English Channel were instigated. We cleared a maximum of 550 k in the process with a normal limit of 500 k. As a speed of 500 knots was reached the airflow over the wing formed its own cloud, which seemed to creep forward and was quite visible through the cockpit side windows. I flew the 1954 SBAC Farnborough Air Show at 500 k, 200 ft, cloud and all. It looked very impressive from the ground but it was damned hard work in the cockpit.

The differences between the Mark I and the Mark II were such that the Mark II's contribution to the development programme was limited once the decision not to proceed with any more Mark II versions was made. This was indeed a tragic decision as it was designed for the low-level role and it was the use of the Mark I in this way later on, when it became clear from reassessment that the effectiveness of enemy defences showed that penetration at high level at subsonic speeds was too dangerous, that caused the demise of the Valiant.

One task assigned to the Mark II was the clearance of rocket-assisted take-offs, utilizing two Super-Sprite Rocket motors attached to the side of the fuselage for use at high/hot airfields. I was amazed when I saw the attachment arrangement. It consisted of a triangular frame that was only held on by one bolt. I remember saying to one of the senior designers, 'You must be joking, it will fall off in no time at all!' and sure enough that is exactly what happened first time out. Immediately after take-off the whole installation, including the rocket motor, fell off before I had time to retract the landing gear. Damage to the wing and particularly to the landing gear housing was significant.

A further problem with the aircraft was the presence of brake judder when applying the brakes. As the use of the aircraft was so limited it did not seem to me that this adverse feature mattered too much. However, another of our test pilots, Dave Glaser, was mesmerized by it and persuaded the design office to request extensive taxi trials to cure it. The inevitable happened when the main landing gear attachment broke under fatigue, causing landing gear collapse, which ended the life of the 'Black Bomber'. It was scrapped in 1958.

By January 1954, production Valiants came into the programme to join the prototype Mark I, which itself continued to play a major role including further assessments by the A & AEE Boscombe Down. The first few production aircraft were assigned to various development uses, so much so that at one time we had more Valiants than Bomber Command itself. Each aircraft had to be passed through its test schedule after arrival from Brooklands with its short 3,600 ft runway to Wisley. During high Mach no. dives on the first production aircraft, as already mentioned, aileron buzz caused the fracture of the aileron rods. Thanks to the extreme skill of Bill Aston, a successful landing was made at Boscombe Down. Thereafter all subsequent aircraft were fitted with steel aileron rods.

One of the development aircraft was involved in a very expensive taxiing accident. Colin Allen determined that the brake accumulators were not charging properly, so he taxied back to the apron via the slight hill down from the runway. Unfortunately, the brakes failed just outside the fire station half way down. He tried to circle the apron while waving a white handkerchief through the cockpit window of his aircraft, but the left wing tip clipped the noses of both the B-29 Washington bombers that were used for Guided Weapons Trials, breaking them off like carrots, before embedding his aircraft in the back of another Valiant which was sticking out of the hangar.

In order to create a training system compatible with a 'wholly new concept' aircraft, RAF Bomber Command set up the operational conversion unit at RAF Gaydon in Warwickshire. The first operational Squadron No. 138, under the Command of Wing Commander Rupert Oakley, was formed in February 1955 before it moved to RAF Wittering in Northamptonshire just a few months later.

Two significant events filled 1956, firstly under Operation Buffalo, Britain's first atom bomb, Blue Danube, was dropped at the Maralinga Range, Australia from Valiant WZ366 piloted by Squadron Leader Ted Flavell. Secondly, and at almost the same time, Valiants from four Squadrons (138, 148, 207 and 214) were sent to Malta to take part in Operation Musketeer during the Suez Campaign. Thus the Valiant became the first postwar four-engined bomber to be used in anger when it dropped conventional high-explosive bombs as opposed to the nuclear weapons for which it had been designed.

In the next few months, preparations were made to prepare an aircraft to drop Britain's first hydrogen bomb from Christmas Island in the Pacific. It was deemed necessary to add special protection against the flash of this device. The whole aeroplane was painted white, removable panels were fitted to all cockpit windows which blotted out any external reference. An escape manoeuvre following release of the weapon was devised. Initially it was thought that a half loop and roll off the top should be made, but it was soon realized that the Valiant did not have enough excess power to complete this manoeuvre successfully without any risk of stalling at the top of the loop. Therefore, I instigated a very tight turn procedure which was used when the single bomb was dropped from 45,000 ft off Malden Island, 400 miles south of Christmas Island on 15 May 1957 by XD818 piloted by Wing Commander Ken Hubbard of No. 49 Squadron. Further weapons were dropped between May and November 1958.

This was an extremely interesting phase in the Valiant programme as I was so heavily involved in its operational role. Valiants achieved high placing in the US Strategic Air Command's bombing competition in Operation Long Shot in Florida and in Operation Iron Horse in California and their reputation was by now worldwide.

I then started on the operational development of in-flight refuelling using the hose and drogue system produced by Flight Refuelling. The Americans used the flying boom system, which required the presence of a boom operator in the tail of the tanker. I did some familiarization flying in Meteors and Canberras at Flight Refuelling's base at Tarrant Rushton near Winborne, Dorset and all seemed fairly straightforward until I flew a Valiant with Pat Hornidge of Flight Refuelling as co-pilot behind a Canberra tanker. Each time I approached ready to insert the receiver's probe into the tanker, the drogue drifted sideways like a startled maiden. We realized that the probe was not long enough and that the strong bow wave from the Valiant was different to that from the smaller aircraft used hitherto. Once the probe was doubled in length, we commenced a programme using two Valiants.

In order to make a successful contact between the drogue and the probe it was necessary to have a closing speed of about 1 knot per second. If the contact was too soft, the fuel coupling did not mate properly, permitting fuel spillage. A contact that was too fast usually led to the hose whipping and removing the end of the probe. The hose drum unit was designed to accept this modest closing speed with a small margin. Once contact was made, the receiver pushed in closer to the tanker. In the case of Valiant to Valiant, the hose drum unit carried an 80 ft hose (100 ft originally) which was wound in to about 60 ft for the correct refuelling position. Accurate flying on the part of the tanker pilot was vital and was not as easy as it may appear because the receiver tended to tip the tanker nose down as it approached. It was the use of two large aircraft together that fully revealed the problems, but after a bit of practice I found that I could make contact quite easily but the characteristics of the drogue were very critical. Initially the drag on the drogue was too high and caused excessive buffeting on the receiver after contact, but when it was reduced contact was almost impossible. After a great deal of trial and error a compromised collapsible drogue was developed for use in the RAF.

The final phase of clearing Valiant to Valiant in-flight refuelling was carried out jointly with A & AEE Boscombe Down in order to save duplication of effort and time. Valiant WZ376 had been modified to carry the hose drum in the bomb bay and was the tanker, while WZ390 was used as a receiver. On one occasion Bill Aston in the tanker and myself in the receiver gave a demonstration while remaining in contact at a flying display at RAF Honington.

The final clearance test flight for Valiant to Valiant was laid down by A & AEE Boscombe Down to consist of the receiver flown by myself and Wing Commander Clive Saxelby OC 'B' Squadron to take off at maximum weight and fly for about ten hours by which time the aircraft was down to a low fuel state. The plan was then to take on a full fuel transfer back to maximum weight at night using the emergency low-rate transfer from the tanker. This configuration required contact to be maintained for approximately twenty-eight minutes. I managed this but it was very demanding in the dark, when one realizes that the receiver nose was about 6–7 ft below the tanker tail. The idea was to fly for another eight hours but weather conditions at Boscombe Down for landing began to deteriorate rapidly so it was agreed to curtail the flight and not have to face a diversion to Kinloss in Scotland. I found twelve hours strapped to a hard ejector seat very uncomfortable and I was extremely stiff when we disembarked.

This was not the end of flight refuelling for me. My close relationship with 'B' Squadron resulted in Wing Commander Saxelby asking me to fly in a Victor bomber which was beginning its in-flight refuelling behind the Valiant tanker. This threw up the same problem as I had experienced on the Valiant, namely the probe was too short and had to be doubled in length. I went through the same exercise in a Vulcan. My next task was to go to RAF Marham and train the RAF Valiant pilots in this new concept which has played such a vital role in RAF operations since 1958.

The Valiant tanker WZ376 continued to be used to clear a number of other aircraft types. When it came to the Lightning, I suggested to Roly Beamont that I would land the tanker at Warton for a pre-flight briefing but he did not consider this to be necessary as we had discussed it all before. However, he set up a closing speed that was much too high, clouted the drogue which caused the hose to whip and smartly removed the end of his probe. I remember Roly's inevitable message coming over the RT: 'I seem to have lost my probe . . .'.

The last Valiant flew out of Brooklands on 27 August 1957 and was delivered to the RAF in September 1957. A total of 104 production aircraft were built, as well as three prototypes. Production rate peaked at four per month and from the sixth delivery in 1955 each aircraft was delivered on schedule or ahead of time, all from the Weybridge assembly line. The operational value of the Valiant was summed up by Air Chief Marshal Sir Kenneth Cross, Commander-in-Chief RAF Bomber Command:

> The Valiant was the backbone of Bomber Command. For two years it was the only aircraft in the Command capable of carrying a nuclear weapon, all the trials which proved these weapons were carried out by Valiants. The complicated electronic navigation and bombing equipment used in all V-Bombers was first tested and tried in Valiant aircraft. Because of its rugged construction and admirable handling qualities it frequently operated from short, sub-standard runways overseas in conditions of high temperature. In bombing competitions against crews of the US Strategic Air Command and in Bomber Command, the Valiant Squadrons covered themselves with glory and convincingly demonstrated the technical proficiency and strike potential of the Royal Air Force. The Valiant was the first British aircraft to be used as a tanker and pioneered flight refuelling in Bomber Command. These are some of the great achievements of this remarkable aircraft. To all those who designed, built and tested it and to those who operated the aeroplane, great credit is due.

Sir George Edwards with his lifelong devotion to the RAF could not have wished for a better tribute – the more so because it also recognized the part played by his team, which he always claimed was the best in the business. I unreservedly admit my own pride in being involved to such an extent.

Valiants equipped nine bomber and tanker squadrons (7, 18, 49, 138, 199, 207 and 214) and one photo-reconnaissance squadron (543), as well as the Operational Conversion Unit and the Bomber Command Development Unit. The main bases from where they operated were Honington, Finingley, Wittering, Marham, Wyton and

Gaydon. Following use in the low-level role for which the aircraft had not been designed, fatigue cracks were found in the front and rear wing spars. The cost of repair was too great to contemplate and the whole Valiant fleet was prematurely scrapped in 1964.

SEVEN

Other Activities

Variety is often said to be the spice of life, and there was plenty of variety apart from the main aircraft projects with which I was involved.

As early as 1940, Dr Barnes Wallis wrote a paper 'A note on a Method of Attacking the Axis Powers'. This note advocated the use of the 'big bomb', 10,000 lb. Resulting from the work of Barnes Wallis came the special bouncing bomb used to break the Möhne and Eder Dams in Germany, which was followed by Tall Boy used against the battleship *Tirpitz* and then by Grand Slam.

It therefore came as no surprise when Vickers entered the guided weapons field although the company's approach was initially cautious. It began at Weybridge with an air-to-ground weapon called Blue Boar which was guided by television. My involvement started when a Lincoln bomber was delivered to Wisley in October 1950. As I was already qualified on the Lincoln, successor to the Lancaster, I was the obvious choice to fly a series of flights dropping Blue Boar on Salisbury Plain. Blue Boar started as a private venture but was taken on by the Government for a few years until it considered that they had an alternative weapon which would work blind and the project was cancelled.

From a test flying point of view, the expansion of guided weapon work led to the build up of a significant fleet of aircraft. Much of the work was done in conjunction with EMI, a company I had always thought made gramophone records. A specially equipped Valetta led the parade and was mainly manned by test engineers from EMI. Soon after, a Canberra B-2 twin-engined bomber arrived from English Electric, Warton; Roly Beamont gave me a short brief on the cockpit on arrival, together with a few pages of scant, type-written pilot's notes. Mutt Summers immediately suggested that I flew the new acquisition and I went on to do a tremendous amount of flying in this aircraft, which was once again for the EMI's benefit. EMI produced a navigator/test engineer named Don Bowen, who later joined Vickers and subsequently flew with Roly Beamont in the TSR2 prototype after the British Aircraft Corporation (BAC) was formed.

Another Canberra followed for use on the Red Dean project, which was an air-to-air weapon. Red Dean became a very large project and a team of nearly 800 was formed to develop it on behalf of the Government. Philip Murphy did most of the flying and encountered unexpected aerobatics when only one Red Dean carried at each wing tip, separated properly on the Aberporth range. The next arrivals were two Meteor fighter aircraft followed finally by two B-29 Bombers assigned from RAF Marham, forming quite some fleet in conjunction with the seven to eight Valiants assigned to various development tasks. It was the most valuable experience for myself and some of our other test pilots and we made the most of it. In fact, industry test pilots in the 1950s were encouraged to fly as many other types of aircraft as possible and it enabled me to have a very short flight in the flying wing AW52 built by Armstrong Whitworth and operated by Aero Flight of the Royal Aircraft Establishment. Unfortunately, a failure of the hydraulic system hastened my return to terra firma, as I had no wish to be left without any form of control.

Red Dean was subsequently cancelled when the Government cancelled the supersonic bomber that would have carried it. This was of course during the era of Duncan

Sandys's White Paper on Britain's Defence which made such a vital impact on the nation's defence policy.

Another diversity occurred when the company bought a Dove, a small passenger aircraft. Initially it was kept under wraps by the sales department but as a great concession, George Lowdell, the Chief Production Test Pilot for Vickers-Armstrongs, was allowed to fly it. I was having a Sunday free for the first time for several weeks as we flew development aircraft seven days a week until noise resulted in a midday Saturday termination. On this particular Sunday I got a telephone call from George Lowdell to say his wife was ill and could I do a trip to Paris to collect some passengers who had been attending the Paris airshow. I had little option other than to agree although I had never flown the Dove before. The flight inspector who always flew with the aircraft showed me the knobs and tits, after which we took off for Villacoublay just outside Paris, where I met the passengers, one of whom was decidedly aggressive. On the way back to Southampton Eastleigh airport he slept off the cause of his aggressiveness and it was preferable not to wake him up. Unfortunately, the grass airfield at Eastleigh was decidedly rough, which made the Dove behave like a bucking bronco. The Commander, as he was, woke up and leapt out of his seat but his crotch was thrown back on to the armrest, and his agony was heard across the airfield. Once inside the customs shed, his mood was not improved as he fell foul of the customs officer. Obviously, not a good trip for the Commander, who smashed two bottles of gin on the floor of the customs shed. Hopefully he was returned to selling battleships in South America.

In 1956 we had a visit from Don Finlay, Chief Initial Projects, Boeing, Seattle. Don was constantly sounding the praises of the Boeing 707 prototype, known as the Dash 80, when his remarks prompted Jock Bryce to say, 'If it is that good, why don't you ask one of us over to fly it?' 'What time does your flight leave?' was Don's response. Much to my delight Jock turned to me and said, 'You go . . .'. So I set off for the United States in a Pan American Stratocruiser which took twelve and a half hours to New York from Shannon airport, our intermediate refuelling stop. I was accompanied by Basil Stephenson from our design office who was obviously interested in the design features of the B-707. We eventually reached Seattle two days later and reported to the Boeing Flight Centre on Boeing Field.

The 707 was airborne when we arrived but returned to Boeing Field to pick us up. I was fascinated when the pilot 'Tex' Johnston, Chief of Boeing Flight Test, ran across the parking apron wearing cowboy boots and a sombrero hat to beckon us aboard. Pointing to the left-hand seat he said 'Sit down Captain' – as far as 'Tex' was concerned any Englishman worked for BOAC. I took an immediate liking to 'Tex', who may have had his wilder moments on occasions, but was a highly professional test pilot when he got down to business. Within a couple of minutes, we were rushing down the runway for a memorable flight covering high speed, stalling and general handling tests with two extra landings. I found the 707 very easy to fly, although one had to be careful on landing not to land with a wing down and risk rubbing an engine pod, and its stalling characteristics were immaculate. I was quite pleased with my performance and my landings were rather better than another Boeing test pilot who was being checked out at the same time. The flight lasted two hours, and the next day I had another two-hour flight, this time with the Project Pilot Jim Gannett. This started a long and valued friendship with 'Tex' Johnston, who will go down as one of the great characters in aviation history. The following year 'Tex' persuaded me to give a lecture on Viscount Icing Trials at the Society of Experimental Test Pilots' Symposium which was part of the World Congress of Flight Conference in Las Vegas.

In 1958 I spent a much longer time with Boeing, Lockheed and Douglas. While in Seattle, I had six more flights in the Dash 80 and 1st production aircraft, as well as security clearance for one flight in a KC135 tanker and one flight in a B-52 bomber. I formed a number of close friendships at Boeing with Dix Loesch, Jim Gannett, Jack Waddell, Harley Beard, Lew Wallick and Brian Wygle, to mention a few. At Lockheed I met Tony Le Vier and Herman 'Fish' Salmon. The visit to Douglas included a trip to Edwards Air Force Base, where I started a lasting friendship with George Jansen, who sadly died in 1994.

I became a member of SETP in 1957, later becoming an Associate Fellow, and finally by being honoured with a Fellowship in 1966. I was in good company as the other Fellows included George Bright, James Gannett of Boeing, Bill Ross of McDonnell Aircraft Corporations (later McDonnell-Douglas), Lt-Col Robert Rushworth (USAF), André Turcat (Sud-Aviation) and myself. Two honorary Fellows were appointed, Sir Frank Whittle and Mr Waldo Waterman. This valuable organization provided many opportunities to learn more and more and also to make contributions on other occasions. I received an even greater honour in 1971 when the coveted Ivan C. Kinchloe Award was made to André Turcat and myself 'for accomplishment of the first flights of Concorde supersonic aircraft'.

SETP blossomed to become a truly international body forming additional sections throughout the USA and adding a European Section, which changes from country to country within Europe each year. I was Chairman of the European Section in 1973 and I fulfilled other duties at SETP Headquarters being Vice-President for a year and poolside reception chairman at the Annual Banquet at the Beverly Hilton Hotel, Beverly Hills, California on two occasions. The latter was not an arduous duty, as it only required the chairman to greet all the attendees and ensure that the lifeguard was on duty to fish out all those who fell in for one reason or another. Being Vice-President did not work out, as the Society's business meetings were held at Lancaster, California and it was just too far away. But I did manage to attend the Annual Symposium and Awards Banquet which was always held at the Beverly Hilton almost every year until 1984. The symposium consisted of two and a half days of technical presentations by test pilots on their work, providing they were within their companies' proprietary rights, and one major visit usually to Edwards Air Force Base or Plant 42, Palmdale. Like so many of these occasions, it was meeting people and talking to them that made it all so worthwhile. The only problem for me was that there was never enough time to do all the things that one wanted. My wife Yvonne used to accompany me to the SETP Symposiums in Los Angeles. In fact she became quite an expert on aeronautical matters by sitting through technical sessions, interspersed by visits to the excellent shopping centres.

In 1956, the Supermarine flight test team joined us at Wisley from Chilboton, near Stockbridge, Hampshire. Mike Lithgow, Dave Morgan and Chunky Horne were involved in the Swift fighter programme and then the Scimitar Naval Fighter. The Swift suffered violent 'pitch up' problems and went into the RAF at numbers significantly below what had been hoped and was beaten into second place by the Hunter. I had my first supersonic experience in a Swift soon after they arrived. This had to be achieved in a fairly steep dive, so I climbed up to about 38,000 ft, rolled upside down and pulled the stick back; this did the trick.

I flew the Scimitar rather more than the Swift. It was a powerful brute with two Rolls-Royce engines. I took part in the 'buddy-buddy' flight refuelling trials but nearly came a cropper carrying out low-level tests at 600 knots over the sea when I ran into thick haze and became completely disorientated as I approached the Isle of Wight. Fortunately, luck was on my side that day and I managed to gain some height into clearer air.

Our communication fleet had been increased over the sole Dove with two Herons and one Beagle 206S. The 1950s certainly provided me with a wide variety of activities and they were very enjoyable and interesting times.

EIGHT

The Vickers Vanguard

The activities detailed in the previous chapter did not mention my main undertaking during 1958 and 1959, which was to address the latest aircraft from the Weybridge stable, namely the Vanguard. The Vickers' Vanguard came into being as a Viscount replacement that would, while retaining the economics of the turbo-prop, provide faster speeds, a greater passenger and freight capacity and a reduction in operating costs.

Discussions between Vickers and BEA started in 1952 and by 1953 potential specifications were being exchanged. The situation was complicated by the fact that Trans Canada (later Air Canada) were thinking along similar lines but the respective solutions differed considerably at the beginning. The success of the Viscount promoted the interest in going to Vickers in the first instance. There was a great deal of debate as to the positioning of the wing, BEA being keen on a high wing, but TCA finding this unacceptable because of the difficulty of de-icing a high wing. The capacity specified for the cargo holds was large and again a high wing had benefits but would have meant a very long landing gear which neither BEA or TCA wanted.

The requirements for range were different as well but in the end acceptable solutions were found to suit both BEA and TCA. The BEA aircraft emerged as the Type V951 and TCA as the Type V952. The gross weight of the V951 was 135,000 lb while the V952 was 6,000 lb heavier at 141,000 lb. A low wing and double bubble fuselage were selected and both types looked almost identical with only very small differences such as the number of cockpit windows (thirteen to twelve). The Canadian order for twenty-three aircraft was won against fierce competition from Lockheed with the Electra. BEA changed their minds once they learnt of the V952 specification and ordered fourteen V953 and six V951 instead of twenty V951 in order to avail themselves of the greater capacity of the V952.

This produced an aircraft that was specially tailored to BEA and TCA requirements but unfortunately this did not result in it being attractive to other customers. Exhaustive sales efforts failed to sell any more new aircraft, but later when the Vanguard appeared on the second-hand market, the Type V952 in particular found its way into a number of airlines around the world.

The Vanguard was powered by four Rolls-Royce Tyne engines, which was a new engine. The combination of a new airframe and a new engine is something that should be avoided where possible – one set of teething problems are enough to deal with, let alone two, but in this case it was unavoidable. Two Elizabethan aircraft (Airspeed Ambassador) loaned from BEA were equipped with one Tyne engine in place of one of its piston engines. There was therefore some in flight experience with the engine but the prototype was beset with problems before first flight when two bearings seized in one of the Tyne engines during ground runs. All four engines were returned to Rolls-Royce immediately and investigation showed that a blocked oil line was the cause of the problem. This of course delayed the first flight for several weeks.

The size and complexity of the task of moving the aircraft by road from Weybridge, where all the Vanguards were built, to Wisley were too great to regard as being practical. Consequently, Jock Bryce as Chief Test Pilot with my collusion agreed to make the first flight from the very short runway at Brooklands which was only 3,600 yards long. We made this flight on 20 January 1959 late in the afternoon. Some reasonably fast taxiing on the runway was possible to ensure that the nose would lift at the designated speed, fuel load was kept to a minimum and the take-off weight was further restricted by not carrying any other crew members. The first take-off went very smoothly except that the throttles crept back a bit once we were airborne. Fortunately, I had my hand on the throttles which enabled me to reapply full power very quickly.

The original intention of the first flight was merely to fly the aircraft from Weybridge to Wisley, which normally took around two minutes and forty seconds. However, everything felt good and so the maiden flight was extended to twenty minutes. During the approach for landing we noted significant flap buffet, but no one took much notice of this until one of them fell off later on. After landing, I selected reverse pitch on the propellers on Jock's command which caused the ailerons to thrash quite violently and nearly broke his wrist. We then showed off a bit by taxiing backwards for a short distance.

I read in one write-up on the Vanguard that the test flying programme went very smoothly. This is what I call typical sales talk. The truth is that we encountered a number of problems during the flight development. First of all, vibration levels in the passenger cabin were not good, especially towards the top of the climb and during cruise. The standards set by the Viscount were simply not present and, as to the stalling characteristics, the Vanguard was a real pig. An extensive programme of fitting wing fences, spoiler strips on the leading edge of the wing and vortex generators were all tried to overcome some very violent tendencies to roll, sometimes inverted, at the stall.

What was required was a sharp nose drop with a minimum amount of roll. It took over 2,000 stalls on the prototype, the majority of which were done by myself, before an acceptable solution was found. It was stalls in turning flight that gave most trouble. A great deal of hard work went into the stalling programme and the changes to wing configurations were made very quickly. After one such change, I taxied out in such a hurry to find out if it worked or not, that I clouted the left wing tip on one of the hangar door supports causing considerable damage. I was frightfully upset at the time but a sympathetic view was taken by my bosses and a new wing tip was readily exchanged and we were back in business within a day or so.

Another interesting feature of the Vanguard was the relatively high-speed envelope. Demonstration of this was made by flying down the design speed diving line (VD). At about 12,000 ft the design limit was some 400 k. I remember achieving this during the structural tests and found it quite amazing. The aircraft itself handled beautifully and attaining the speed was no big deal but the noise in the cockpit was almost unbelievable. This was not too surprising when one pictures this great tub of an aircraft thundering along at 400 k.

The first production aircraft for BEA appeared quite quickly and flew in April 1959 and the second one in July 1959. These were the days when Vickers really knew how to turn out aircraft. The second production aircraft was scheduled for tropical and high-altitude trials in Khartoum and Johannesburg which were originally going to be done by Jock Bryce. However, it became necessary for him to take the first BEA aircraft to Canada so that TCA could examine it. On the way to Montreal, it was observed that the flight-control locks were moving towards the locked position as a result of encountering very cold ambient temperatures. A cure for this potentially lethal problem was found very quickly.

Accompanying Jock Bryce with the technical team for the tropical trials was Joe Leach, the Flight Test Manager. Jock was senior to Joe Leach in the pecking order and was therefore the Tour Manager, responsible for the whole party. When I took on the trip from Jock this caused some concern to Joe who demanded that he should be Tour Manager. I remember saying that I did not mind what he was, providing that it was clearly understood that once the aircraft moved under its own power, I was in charge. I had always got on well with Joe and did so thereafter but this particular situation seemed to worry him. He decided that he must hold the aircraft Flight Manual and be first down the steps of the aircraft. However, when we arrived at our first stop, Rome, things did not quite turn out as he had envisaged. The usual reception committee of customs officers, immigration and other officials were standing at the bottom steps asking for the Captain. Joe duly introduced himself as the Tour Manager, to which one of them responded in Italian 'I don't give a bugger who you are, I want the Captain!' I then produced myself to their satisfaction and held the Flight Manual thereafter.

We spent ten days in Khartoum which was a place I had never liked. The Grand Hotel itself was full, so we had the annex which was an old Nile steamboat without the comfort of air conditioning. We then moved to Johannesburg for thirty-eight days where the tests were mainly measuring take-off and landing distance and a great deal of climb performance. The cockpit was very hot and I remember losing nearly 7 lb in one day. Due to finger trouble on the part of co-pilot, Dickie Rymer, who had joined us at the outset of the Viscount programme, and myself, we burst all the tyres on one landing, which apart from making the runway look like a battlefield, caused damage to hydraulic pipes on the landing gear and also to the landing gear doors. This held us up for a few days while spares were flown out from the UK.

Before leaving Wisley, I had made it quite clear that I was not prepared to use the engine power off technique as used previously for measured landings. Instead a technique very close to a normal approach was adopted. As one was entirely in control of the situation, I do not believe that the distances measured suffered any penalty. One critical test during measured landings was demonstrating a landing at the target threshold speed minus 5 k on the extended forward centre of gravity limit. I accomplished this manoeuvre in front of the two representatives from the Air Registration Board and, feeling thoroughly relieved, I turned around and said, 'Okay?', to which they responded 'No. Do it again!' This made me rather angry until I realized that the ARB pilot did not want to do one himself. The reason why this test was difficult was due to the likelihood of running out of elevator control and thumping the aircraft into the runway.

The tests in Johannesburg lasted until 6 December 1959 when we flew from Johannesburg to Salisbury, where we made two test flights and one demonstration on 7 December. On 8 December we went to Nairobi for a demonstration flight with three test flights on 9 and 10 December. On 11 December we had quite a day starting with Nairobi to Khartoum and Khartoum to Cairo. On arrival at Cairo one of the sales people announced that two or perhaps three demonstration flights were planned. He could see that I was not too pleased because firstly we were tired and secondly we were due in Beirut that night. He kept saying 'Never mind, I have a nice lunch laid on for you', and I had noticed several large turkeys laid out on the buffet table with oranges stuffed into every orifice, including one up the ass. We managed to do the turkeys justice before completing the programme and leaving for Beirut by night. I literally belted the Vanguard to Beirut and did the trip in one hour and five minutes flying at 1,000 ft at over 300 k Indicated Air Speed.

On arrival at Beirut, I found that we were expected to attend a high-powered dinner. The first course of sheep's eyes was too much for me and I retired hurt and this was just as well because the 12 December 1959 proved to be even more dramatic. On the second of two demonstration flights with Dickie Rymer in the left-hand seats the landing gear failed to extend. The left landing gear remained firmly up and all attempts to dislodge it by applying loads and pulling up violently failed to have any effect. Eventually I went underneath the floor boards and started pulling wires and pushing rods. Sometimes it was obvious to Dickie that I was pulling the wrong thing as he found the elevator trimmer moving on its own. Just as I emerged from a hole in the floor, the same sales department member with his turkeys in Cairo said to me, 'Do you think that it is time that we went back to Beirut?' I said, 'What the hell do you think that we have been doing for the last fifty minutes?!' Miraculously the left landing gear did come down but I would have stuffed the salesmen if I had a suitable orange available. We found that a bolt had fouled the landing gear door and so caused the problem, which was fixed by applying a very smooth lining material to the inside of the door. Fortunately, the rest of the trip Beirut to Nice, Nice to Gatwick and Gatwick to Wisley was uneventful.

Normal development flying aimed towards certification continued in the early part of 1960 and the first TCA aircraft joined the flight test programme on 21 May 1960. The programme then suffered a major hiccup. During a BEA proving flight between London and Athens, one of the Tyne engines had to be shut down. A bench test on the same engine reproduced the same fault, while a second bench test wrecked the engine test house when a compressor failed. Rolls-Royce immediately recommended grounding all Vanguards until they had identified the problem, which was found to be incorrect heat treatment on the forgings in the compressor discs. This following numerous engine changes during the first ninety hours on the prototype in 1957 which came to twelve did not make the Tyne engine a big favourite of mine.

The grounding delayed the planned entry date for commercial service with BEA schedule for 1 July 1969, leaving BEA very short of capacity. Modified Tyne engines appeared in the autumn for an intense period of test flights, with sufficient evidence being obtained to satisfy the Air Registration Board, who granted a Certificate of Airworthiness on 2 December 1960. BEA had to undertake some crew refresher training but still managed to start commercial services over the Christmas period. TCA schedules for operations were not quite so tight and were met when the inaugural commercial flight took place on 1 February 1961 as originally planned from Montreal to Vancouver via Toronto, Winnipeg, Regina and Calgary.

The Vanguard, like most Vickers products, had many attributes but the timing of the aircraft was wrong as far as a number of airlines were concerned, who saw it in the minds of the travelling public as a somewhat unfashionable and slow turbo-prop when compared to jet airliners like the Caravelle. Nevertheless, both BEA and TCA did pretty well with the Vanguard. BEA worked its Vanguard fleet very hard and recorded high utilizations of nearly eighty hours per week per aircraft. Across the Atlantic, TCA had received its initial order for twenty aircraft by the end of 1961 and as more aircraft arrived TCA expanded their use covering not only trans-Canada routes but also high density links between Montreal and Toronto. It was also used on services from Montreal and Toronto to the United States and to the West Indies. It was also employed on eastern seaboard routes including Nova Scotia and Newfoundland. TCA received two of its remaining three aircraft in 1962, but the last delivery was not made until 3 April 1964.

Although the time advantage of the jet airliners was very small over the shorter routes, the passenger appeal of the jet gradually pushed the Vanguard into second place. BEA began replacing the Vanguard on prestige routes such as London to Paris with Comets and Tridents in the face of Caravelle competition.

The advent of the BAC1-11 on UK domestic routes by British United and British Eagle put a further nail in the coffin. Rolls-Royce experienced further trouble with the Tyne engine in the form of fatigue on the high-pressure compression chamber. Once again the problems were solved by fast work from Rolls-Royce. But 1966 saw the use of the Vanguard as a freighter within BEA on a limited basis. By 1968 full conversion to freighter standard began. The conversion required the fitting of a large freight door 139 in wide and 80 in high to the forward left side of the fuselage. The floor was strengthened, a roller system installed on it and the cabin windows blanked out. BEA renamed the freighter Merchantman.

For some years BEA (which became British Airways on 1 April 1974) operated a mixed fleet of Vanguards and Merchantman. The role of the passenger aircraft continued to decline and more aircraft were converted to Merchantman standard. Gradually other operators come into the picture as more aircraft ex-BEA and ex-TCA appeared on the second-hand market. The history of how and where the forty-odd Vanguard/Merchantman finished up is interesting – the final use of the Merchantman in the 1990s was carrying cattle and then racehorses, for which it was eminently suitable. When the British Aircraft Corporation was formed in 1960, the Vanguard remained on the Vickers account and it was therefore Vickers who had to stand the financial loss on this programme, which had to be regarded as commercially unsuccessful. I had become quite attached to Vanguard but I did no real flying on the type after June 1962, when the VC10 flew for the first time and filled my life for several years.

NINE

The British Aircraft Corporation and the Vickers-Armstrongs VC10/Super VC10

Before moving on to my next project, the VC10/ Super VC10, it is relevant to mention the formation of the British Aircraft Corporation. It is a fact that at the end of the Second World War there were twenty-seven British airframe companies and eight aircraft engine companies; rationalization was therefore inevitable. Although this was recognized in the late 1940s, ten years of wasted opportunity followed and it was during this time that Duncan Sandys's White Paper of 1957 effectively killed off all advanced military projects like the Avro 730 supersonic Mach=2+ bomber, the Fairey Delta fighter (OR 329), the Saunders-Roe 177 rocket/jet fighter, the Hawker P1083 and the Hawker P1121 Mach=2+. Only the P1 Lightning and the Canberra replacement (OR 339) were left untouched. So there was not much to play for. The opportunity to produce the first big jet transport for transatlantic operations went with the Vickers V1000 military transport and its civil counterpart. The requirement for a jet for BEA was another fiasco, the Vickers VC11 being considered too big, although both the VC11 and de Havilland 121 (Trident) powered by Rolls-Royce Medway engine would have been world class and probably seen off the Boeing 727. The Medway was never built in production, which cost the true potential of the BAC1-11 later on. BEA tailored the de Havilland 121

(Trident) to their own specific requirements by cutting down its seating and reducing its range, which condemned it to the Rolls-Royce Spey engine which had limited development potential. The BEA order was won by a consortium of de Havilland (67½%), Hunting (22½%) and Fairey (10%).

By 1958, the Government tabled a policy encouraging the industry to re-shape itself in such a manner that civil projects at least could be financed without Government assistance. There were many rumours of various marriages taking place during this period with the intention of winning OR 399. When the Conservative Government won a further term of office in 1959, Duncan Sandys, who became Minister of Aviation and a member of the Cabinet, following the amalgamation of the Ministry of Supply and the civil side of the Ministry of Transport into a single Ministry of Aviation, nearly achieved his aim. Three groups were formed not two because neither of the airframe groups wanted helicopters, which were all put into Westland Aircraft Limited with Hawker Siddeley Aviation and British Aircraft Corporation as the two main airframe groups.

Charles Gardner's book *British Aircraft Corporation* details how Bristol and Hunting Aircraft joined with Vickers and English Electric to create British Aircraft Corporation (BAC). By the time BAC was formed, the VC10 was well under way. The immediate effect of the new organization on myself was that Jock Bryce was appointed Chief Test Pilot of BAC with Roly Beamont as his deputy, while I became Chief Test Pilot of Vickers-Armstrongs' (Weybridge) Division. It was, however, made clear to me that Jock would wish to retain the right to command the VC10 and the BAC1-11 on their first flights, even though they were both the responsibility of the Weybridge Division. I thought at the time this was a rather poorly conceived arrangement and my opinion never changed especially during the times when no one knew who was supposed to be doing what. Fortunately the friendship and working relationship between Jock and myself that had evolved over the years made the situation possible to survive without too many sleepless nights.

Much has already been written about the political saga behind the VC10 and the different attitudes of various BOAC Chairmen. The facts as I saw them were that BOAC got what they asked for. It was BOAC who chose to come to Vickers with a specification calling for a payload of around 35,000 lb to be carried over a 2,500 mile sector on the airline's Eastern and African routes to Australia. This entailed operation in high ambient temperatures, high altitude, with relatively short runways and on difficult route sectors involving strong head winds. The critical airports were Kano (Nigeria), Nairobi (Kenya) and Johannesburg (South Africa) and the sectors were Singapore to Karachi and Kano to London. This was a tough assignment and the need to be fully competitive in every respect with the new American jets, which were likely to set the trend of international air travel, was fully recognized. Special attention to passenger appeal was also required in the form of reduced cabin noise levels and improved air conditioning.

I distinctly remember the in-depth discussions and meetings with BOAC which defined the detailed design of the aircraft systems. It was abundantly clear that a rear-engined configuration was superior in providing the exceptional performance required to meet BOAC's route requirements. The uncluttered wing became a vital part in achieving the primary design objective of airfield performance. The penalty for this layout was a higher basic airframe weight due to lack of bending relief afforded by wing-mounted engines. However, better lifting characteristics would be obtained from this clean wing configuration. The advantages claimed for rear-mounted engines were greatly improved airfield performance with higher payload uplift and thus better operating economy; greater versatility on the routes; improved control characteristics and lower approved approach speeds; a new standard of cabin noise levels and vibration; reduced fire hazards in wheels up landing; excellent ditching characteristics; reduced risk of structural damage from the effects of jet efflux. All this convinced Vickers and BOAC that the proposed layout was best suited to meet the demanding specification. In fact other manufacturers and operators adopted a similar philosophy for the Boeing 727, Trident, BAC1-11, DC9, Tupolev 134/154, Fokker 28/100 and Ilyushin 62.

It was ultimately found that relatively small changes would give the aircraft a transatlantic capability. These changes were incorporated in the basic VC10 design during 1957 and BOAC's original order was announced in January 1958. Having decided on the clean wing arrangement, the opportunity was taken to provide an advanced wing section and shape. The VC10 thus became the first aircraft to use the supercritical peaky wing design theory. A great deal of work was required to turn this theory into data to design the VC10 wing. To aid meeting the stringent airfield requirements, the clean wing was able to incorporate very efficient leading edge slats and very powerful, large area 65 per cent span Fowler flaps in the wing trailing edge. Unfortunately, the VC10 wing was not right first time and a 4 per cent leading edge extension and some re-shaping was required to reduce the drag which was higher than estimated. The incidence of the engine nacelles was changed and a beaver tail fairing between the rear nacelles was fitted in order to reduce engine interference and installation drag effects. The Douglas DC8 used a similar peaky wing design and also suffered drag problems.

Special attention was paid to the passenger-cabin appointments overseen by specialist industrial designers (Charles Butler Associates of New York). The main cabin and service-access doors incorporated an ingenious outward-opening plug mechanism. A 30 ton Freon vapour cycle cabin refrigeration system was installed in the wing root leading edge for ground and flight use in tropical conditions in order to prevent passenger discomfort.

A split system philosophy was adopted for all the major systems. Both half systems were arranged to operate simultaneously and each designed to act as emergency stand by for the other. The principle eliminated the possibility of the emergency system lying dormant with the potential risk of faults, which would only be known when the system was needed. The split system concept was applied to all vital systems – hydraulics, electrics, air conditioning, flying controls and landing aids. The duplex autopilots provided an auto-land capability, which was certificated but not used very much due to maintenance costs.

The primary flying control surfaces were split into segments, each driven by its own electro-hydraulic power unit: 4 ailerons, 4 elevators, 3 rudders and 6 wing-mounted hydraulically operated spoiler/speed linked sections. The movable tailplane was hydraulically operated. In order to cater for four engine flame-out, drop out electrical and hydraulic ram air-driven turbines (Elrat and Hyrat) were fitted. The units were used in anger on one occasion when four-engine flame-out occurred due to fuel mismanagement at 35,000 ft until normal engine power was re-established at 23,000 ft. This occurred on a commercial service and confirmed the claim by the Air Registration Board that all four-engined jet transports had suffered similar occurrences at sometime.

As a consequence of the type of runway from which the aircraft was to operate, with low bearing classifications, the aircraft was fitted with a four-wheel main bogy undercarriage carrying large low-pressure tyres and fully duplicated hydraulically operated wheel brakes.

The flight control system was regarded as the best in the world at the time while in all other areas it was designed as an aerial Rolls-Royce. No doubt this caused some weight penalty. It was a very easy aircraft to fly in the normal flight envelope and was generally loved by the passengers and by the pilots that flew it. A full-sized Integrated Flying Control and Hydraulic Test Rig representing the actual shape of the VC10 was laid out in one of the hangars at Brooklands. Jock Bryce and myself spent many hours on this rig learning the systems, procedures, emergency drills and determining acceptable friction levels. Our preparation for first flight was long and extensive. One funny incident occurred when we were preparing the checklists, which we had to do ourselves, when Jock decided that he wanted to ask me a question on the air conditioning. Consequently, he wrote '5. Air Conditioning – see E.B.T.' Unfortunately, the checklist went off to the printers before I had resolved the question. As a result, the air conditioning was labelled 'See E.B.T.' until the time of the first flight when we stuck a piece of hand-written paper over it.

The runway at Brooklands was lengthened by 400 ft for the first flight but, because of the position of a monument to Lord Brabazon at the north end, it was not possible to re-site the runway in the proper manner. This produced an extension at a slight angle to the main strip and a taxiway was built round Lord Brabazon's monument so that the take-off run could at least start at the beginning of the extension. It meant that at the very low take-off weight for flights out of Brookland, the aircraft came around the corner doing about 100 knots. This half-baked arrangement did permit the raising of the nosewheel to check elevator response and putting it back on the runway in time to stop. This is what was done before first flight.

The first flight of GARTA took place on 29 June 1962 with Jock Bryce, myself and another test pilot Bill Cairns acting as flight engineer. It was going to be Jasper Jarvis, another test pilot, until he decided that having bust his car, he could not come to work and when the firm was not able and willing to supply one, he decided to stay on leave, writing me a note to that effect. When I got it, I sent a car for him and we had a short conversation but nothing would change his mind. As a result I told him to stay on leave and not to come back. Bill Cairns was, therefore, very much thrust on to the scene with only a few a weeks to go. He coped admirably and subsequently became the VC10/Super VC10 project pilot.

The weather for the first flight was a little bit hazy and it was originally intended to make the first landing at Boscombe Down. A weather reconnaissance by one of the communication aircraft reported that visibility looked worse from the ground. This sounded good enough, so we took off without any further delay. The Brooklands runway looked pretty small from inside what was in those days a very large aircraft. Once airborne, the weather report was completely duff and visibility was not good at all and therefore Jock elected to land at Wisley and not Boscombe Down, as we knew the local area like the back of our hands.

The first landing was a real 'greaser' and the quality of the landing gear was very obvious. The flight turned out to be another air lift from Brooklands to Wisley, but gave sufficient time to assess, initially, the excellent flying qualities of the VC10. It felt really good up in the front of Europe's largest ever aircraft in what can only be described as a very spacious cockpit.

Measurement of the VC10's cruise performance was a very high priority. Some of the flying was done from Boscombe Down as Wisley (at 6,600 ft) was not long enough for take-offs at near maximum weight. The initial results immediately showed that we had a major problem with the cruise performance as total drag was several per cent above estimate. Immediate 'fixes' such as improved sealing of the slats and flaps did not provide a solution. Nearly 500 hours of performance flying using mainly aircraft number four GARVE were carried out and significant changes like angle of incidence of the engine nacelles were tried. The type 1101 VC10 never did fully recover its performance and even funny shaped wing tips on Types 1102 (Ghana Airways) and 1103 (British United) only gave small gains. It was not until the Super VC10 wing was developed with some re-shaping and a 4 per cent leading edge extension that the situation improved considerably.

The first production aircraft GARVA followed GARTA in December 1962 with GARVB, GARVC and GARVE all following fairly quickly. GARTA got on with full flight envelope exploration and the flutter clearance. Our Chief Engineer, Hugh Hemsley had a reluctance to finish the flutter programme up to the design limit of M=0.92 for no apparent reason but eventually we got there.

We started exploring the stalling characteristics. The Type 1101 very nearly met the classical nose drop criteria at aft centre of gravity except with landing flap (45°) and the abnormal cases of flaps only, no leading edge slats. In the aircraft clean case the pre-stall buffet was enormously high.

The stalling programme had to be approached very carefully because of the potential of entering what is often referred to as the 'deep stall' or 'stable stall'. The configuration of the T-tail and rear-mounted engines has, in most cases, a marked effect on the stall and post-stall characteristics. At low angles of attack, the low tail contributes less to stability, because of its position in a region of higher downwash behind the wing. As angle of attack increases, the tail moves below the high downwash region and becomes more effective. It remains effective to high angles of attack. The high tail, on the other hand, starts in a region of low downwash at low angles of attack and therefore contributes more stability. As angle of attack approaches the stall angles, the high tail moves into the high downwash region created by the wing and its contribution to stability is reduced. This reduced contribution can be delayed until after the stall angle if the tail is located high enough. Some experts think that the loss of air speed (Q) at the tail is the most significant in this problem. If angle of attack is further increased, the high tail penetrates deeper into the aircraft wake and becomes even less effective. Pitch control effectiveness is also reduced by the wake and this value determines whether or not the aircraft can be recovered from very high angles of attack. Penetration to such angles is a broad definition of the deep stall. Another definition associates the deep stall with a locked-in condition where recovery is impossible. This is what happened to the first BAC1-11, which will be dealt with later.

The in-flight technique for the VC10 was set round the installation of duplicated angle of attack indicators. Based on wind tunnel, I was made aware of the angle of attack where pitch up might occur and I was briefed to stop at a particular angle of attack on each test. A recovery parachute was not fitted, although in retrospect, it probably should have been.

The tests showed that the approach to the stall did not have adequate natural warning (only in the clear case where there was ample pre-stall buffet) and that artificial warning in the form of a duplicated stick shaker was required. Various combinations of wing fences and leading edge disrupters were tried to produce an unmistakable nose drop at the stall. Of these, a deep in board fence at approximately 18 per cent gross semi-span fitted to the top surface of the wing, proved to be the most powerful in removing a slight instability before the stall. The fence caused a small reduction in maximum lift coefficient (CLmax), which did not please the design office but I was not prepared to accept the aircraft without it.

Towards the end of 1963 it was decided to fit a stick pusher, a decision which had already been taken by de Havilland on the Trident, as the only way of showing compliance with the regulations. Vickers' first design of a stick pusher did not push the stick forward but merely inserted a down movement on the elevators. This was rejected by D.P. Davies of the Air Registration Board and so a pneumatic powered device situated under the cockpit floor and below the control column was fitted. It required a pull force of some 80 lb to overcome it and the effect of an inadvertent push had to be demonstrated through the whole speed range including the actual landing flare. Automatic ignition of the engines was also provided. Fuselage-mounted probes supply the signals for the shaker and pusher systems and are set at pre-determined angles of attack for the different flap position and are modified by flap position, slat position and rate of change of angle of attack.

While the ARB were prepared to accept stick pushers, they felt that the substitution of natural characteristics by artificial means required very comprehensive testing in order to demonstrate that such devices would cover all possible exigencies. Consequently, very vicious stalls were conducted, producing rates of change as high as 8° per second, by setting the aircraft in a turn of 30–40° bank, reducing speed slightly and then applying what D.P. Davies described as 'a ****ing great pull'. With aircraft displaying conventional stalling characteristics, it was only necessary to reduce speed at greater than 1 K per second to demonstrate a dynamic stall.

The last day of 1963 nearly brought the stalling programme to an abrupt end. I was just recovering from a clean stall when at about 250 K all hell broke loose as GARTA started shaking violently. There was a shout from the Senior Observer, Chris Mullen, who was looking at the tail through his periscope, 'Right inner elevator'. I was quite certain that GARTA was going to come apart and it nearly did, so I fired the escape hatch door and ordered the crew to bale out. The flight engineer, Roy Mole, could not get out of his seat and the same applied to the co-pilot Captain Peter Cane of BOAC, while the crew in the back could not hear me above the general racket. I managed to reduce speed to about 160 k which put me very close to pre-stall buffet, whereupon the violent vibrations and oscillations calmed down to a smaller amount. The escape chute which went through the front forward hold had collapsed and gone out when the door was jettisoned, so it was as well nobody tried to use it and only a jangled bunch of metal remained. I made a very gentle return towards Wisley under May Day conditions and soon realized that I had lost half the aircraft services. However, the split system principle worked very well but I had to free-fall the right landing gear. After-flight inspection revealed that the two right-hand engines had rotated 2 in and in doing so pulled off hydraulic pipes and air-conditioning pipes. The right inner elevator had broken its attachment bracket which had set up flutter of that surface. Two fin attachment bolts were severed. In fact poor GARTA with whom I had developed a great bond of affection was in a sorry state. I think that we had done about 2,300 stalls together.

This was a bad New Year's Eve as we had already lost the first BAC1-11 and all its crew in October. Sir George Edwards and our Vice-Chairman, Sir Geoffrey Tuttle, came rushing over to me at Wisley from Weybridge. I can remember saying 'If this is 1963 you can keep it. All that is left is for me is to be caught drunk in charge tonight.' Sir Geoffrey immediately said 'You must have a chauffeur for the evening.' As 1964 dawned the only thing I remember was the chauffeur saying at the door to my flat, 'Can you manage the rest of the way on your own Sir?' The next day I received a call from Freddie Laker (now Sir Freddie), my old friend, saying 'I am glad that you did not break your neck yesterday, but how much will that one be going for now?' Anyway GARTA was repaired by early February and the programme continued.

Prior to certification of the standard Type 1101 GARVF (a/c 5) carried out 1,000 hours of route flying by BOAC. Captain Dexter Field and Captain Peter Cane flew with us throughout the whole development programme. Certification took place on 22 April 1964.

I thought the ARB Council Meeting that would approve the granting of a Certificate of Airworthiness would be very straightforward and I had arranged to telephone Dave Davies at the end of it. Dave Davies had been full of praise for the VC10's flying characteristics but there were one or two points which he did not like. The first stick pusher was only approved on a temporary basis until the pneumatic version was retrofitted, the margin between rough air speed and onset of pre-stall buffet at 20,000 ft was rather small and the rate of roll with speed brakes extended at design diving speed was low and only about 7°/second. He painted this picture in such a manner that I think he frightened the ARB Council, who were very conscious of the fact that they were certifying Europe's largest ever aircraft. Consequently, the answer to my question 'Well, Dave . . .' was, 'Well, you just made it!' I needed a couple of stiff ones to simmer down.

The first commercial service commenced on 29 April from London to Lagos using GARVJ under the command of Captain A.S.M. ('Flaps') Rendall (Flight Manager VC10). By this time 75 captains, 127 first officers and 72 flight engineers had completed flight training courses.

The nucleus of the BOAC training captains were trained by Dennis Hayley-Bell and Lew Roberts under the contractual arrangements between BAC and BOAC. The same principle applied to British United, Ghana and East African Airways. During the course of this large amount of flying, Lew Roberts reached the monthly limit of 135 hours as laid down by the Civil Aviation Authority. This required him to have a medical check-up with an approved CAA doctor before continuing. He was examined and passed by a doctor called Tom Ellis Williams who, I learnt later, came from near my parents original home in Llanelli. This started a long and close relationship between all my pilots, including myself and Tom Williams. I persuaded BAC to take him on as our Flight Surgeon and in so doing BAC Weybridge was the first aircraft manufacturer in the UK to do so. Tom

Williams became an integral part of our team and was a wonderful friend to have but that is not to say that he was anything but extremely strict over the medical standards that we had to maintain. Tom Williams looked after all the BAC air crew at Fairford until he sadly died, after which he was replaced by another wonderful individual, Dr Pat Cassidy of Cirencester. Many of us owe these two exceptional doctors a very great deal.

Behind the scenes the usual round of political wranglings had been going on and any word of manufacturers' difficulties was met without any sympathy from Sir Giles Guthrie in particular. The order for thirty-five VC10s was changed to twelve Standards and the remainder to be Super VC10s. Vickers' original proposal was for an aircraft carrying 212 passengers by extending the fuselage by 28 ft and using a 24,000 lb thrust Conway development. This was rejected by BOAC as being too big. Instead a more modest fuselage extension of 13 ft was adopted thus increasing the passenger capacity from 135 to 163. The real crab that BOAC had about the Super VC10 was that it did not carry a proper commercial load non-stop to Los Angeles/San Francisco. Lower seat mile costs were a natural corollary to this still substantial increase in size. Throughout its working life the Super VC10 did a superb job despite a headwind of prejudice which at times, before it entered service, reached gale force.

I made the first flight of the Super VC 10 from Brooklands on 7 May 1964. Most aircraft improve with increase in size and the VC10 was no exception. The Super VC10 handled beautifully and was liked even more than the Standard. Fitted from day one with a stick pusher, problem areas such as stalls were very straightforward. The only area where there might have been a difficulty was rate of roll at design diving speed with speed brakes extended but this was overcome by positioning the air brake selector lever to coincide exactly with the angle of the speed brakes themselves. This was done automatically. Coincident with the Super VC10 programme, I inherited the BAC1-11 programme, especially the stalling programme (see chapter 10).

The development of the Super VC10 was completed in time for the commercial scheduled service to take place on 1 April 1965 from London–New York–San Francisco under the command of Captain Norman Todd. Production aircraft followed at a steady flow along with two Standard aircraft for British United and three for Ghana Airways. British United specified a large freight door, 84 in by 140 in, on the left forward fuselage. This then became a standard feature which was also incorporated on five Super VC10s ordered by East African Airways. These minor changes had little effect on the aircraft and most of the flying was therefore very routine.

An increased order for fourteen VC10s placed by the RAF came as a great morale booster, the initial order for five aircraft having been placed in 1961. The RAF VC10 Type 110 was a combination of the VC10 and Super VC10. The fuselage size was that of VC10 but with a strengthened floor while the wing, fin, landing gear and engines RCO Conway 43 were the same as fitted on the Super VC10. An extra 1,355 gallons of fuel was carried in the wet fin.

The RAF designation for the aircraft was C.Mk1 and in-flight refuelling capability was also incorporated. The flight programme was quite extensive because of these changes and the aircraft was certificated to Civil Standards. It did go to the Aeroplane and Armament Experimental Establishment for a brief service assessment. The first flight of XR 806, made by myself, took place from Brooklands on 26 November 1965, the trials proving to be relatively trouble free and the A & AEE assessment being entirely favourable.

By this time I had already been assigned to the Concorde project. Throughout the C.Mk1 programme we had an RAF crew attached to Vickers under Squadron Leader Brian Taylor. This helped the introduction of the type into No. 10

Squadron, who still operate the aircraft today. This is no mean achievement and an enormous tribute to the RAF and to the manufacturers themselves.

The VC10/Super VC10 story is in some ways a sad one, but the aircraft's record speaks for itself when one considers some of its achievements. During the years 1970–1, sixteen Super VC10s flew 70,347 revenue hours which is 4,397 hours per aircraft per year or just over twelve hours per aircraft over each and every day of the year. The tragedy of the VC10 story is that by the time the operator BOAC and the manufacturer had become good at making big British aircraft earn their keep efficiently, the national decision had been taken to stop making them.

I look back on the VC10 years with a combination of pride and disgust. Pride for being part of the development of a great aeroplane that managed to survive more trials and tribulations than it deserved. Disgust at the political jockeying and the interplay of extreme vacillations of the management, the politics of the airline itself and national political factors far beyond Vickers' influence, let alone control.

I shall never forget that in January 1965 Sir Giles Guthrie became Chairman of BOAC (three months before the first Standard VC10 services were introduced). One of his first drastic reorganizations within BOAC was to cut the order for thirty Super VC10s to seven and then to cancel the lot and buy Boeing 707-320Cs instead. At the same time there was a well-orchestrated campaign against the VC10 in the media. It is hard to sell an aircraft under such circumstances and sales prospects were seriously harmed, notably to Middle East Airlines. It is worth recording that in 1972–5 BOAC reported that Super VC10s were averaging 11.09 hours per day compared to 707s' 8.7 hours and that operating costs per revenue flying hours were £486 for the Super VC10 and the £510 for 707. The Government did not accept the Guthrie All-American plan and ordered BOAC to take seventeen.

Thus the final orders for VC10/SuperVC10 were:

	Standard	Super
BOAC	12	17
British United	3	
Ghana	3	
East Africa		5
RAF	14 (Military)	

GARTA, the prototype, did enter airline service and following a stint with Middle East Airlines finished up in British United where it was unfortunately written off as a result of a heavy landing. The final total was fifty-four aircraft which represented a £20 million loss to Vickers.

A number of interesting possibilities were looked into using twin and even triple fuselage designs with four or even six rear-mounted engines, while another version was a front-loading freighter. However, the most exciting project was a potential order for sixty for the RAF to carry the American Skybolt ballistic missile of which up to four could be installed on the clean wing of the VC10. The British Government's flirtation with Skybolt as a primary nuclear deterrent weapon resulted in the cancellation of UK's own Blue Streak intercontinental ballistic missile but Skybolt was abandoned by the Americans due to technical problems in both propulsion and guidance. In 1962 the British Government adopted the undersea launched American Polaris system instead.

The VC10 was clearly determined not to die without a fight. Once again it was the annual cricket match at Weybridge against the Operational Requirements Branch of the Ministry of Defence. I was talking to Air Marshal Sir Peter Terry when he asked me what I had been doing recently. I mentioned that I had just been out to East Africa to collect one of the four Super VC10s which were being re-possessed as a result of failure to pay for them. This interested him considerably and he said, 'That is just what I may be looking for.'

Following more detailed discussions that evening with Peter Terry, Ernie Marshall and Sir George Edwards, with calculations for conversion of nine VC10/Super VC10 into aerial tankers done on a scrap of paper, I remember a figure of about £35 million being spoken of and a serious project had begun. A contract to convert five Standard VC10s (ex-BOAC/BA and subsequently operated by Gulf Air in the Middle East) and four Super VC10s (ex-East Africa Airways) was placed in May 1979. Although the design authority remained within the Weybridge design office, the actual conversion was planned to be done at Filton; designation for the two types was VC10 KMk2 and VC10 KMk3 respectively. Each tanker had three refuelling units – one mounted in a pod under each outer wing and one in a specially constructed bay in the underside of the rear fuselage.

The idea behind this arrangement was to provide a refuelling capability for two small fighter type aircraft at the same time, while large aircraft would use the centre point position, which provided a higher rate. An additional fuel tank made up of five cells was installed in the passenger cabin above the floor. Access into the Supers was easy, because of the large freight door, but it was necessary to cut a hole in the top of the fuselage on the Standards in order to get the tank in. In order to standardize with the C.Mk1s, the same engines were fitted to the nine tankers as was a Turbomeca 'Artouste' gas turbine auxiliary power unit in the tailcone. Each tanker was fitted with the same 9 ft probe as on the C.Mk1. The carriage of 17/18 passengers was also required.

John Cochrane, Peter Baker and myself, all with previous VC10 experience marshalled the nine aircraft to Filton from Stanstead and other places. All were parked outside the Brabazon as no more than four could fit into the space available, the use of 'Driclad' bags being adopted to protect those parked outside. This proved a nightmare programme. The first job was to conduct a complete inspection of the basic structure for all the obvious possible defects especially corrosion, of which there was plenty.

I soon found myself inheriting the tanker conversion programme as Project Director. I thought that the Filton approach to this programme was lethargic to put it mildly and I had to fight to get the programme moving as other projects always seemed to take higher priority. The storage of parts taken off the various aircraft was abysmal and a large amount were lost or damaged. In the end when I became Director and General Manager of Filton I got the programme really cracking but at a price, for which I was heavily criticized. The production man-hours were much higher than budget and the cost overran, causing a considerable row within the Ministry of Defence. The original £35 million tidied up to about £60 million, later the figure reaching nearly £120 million.

In retrospect, I suppose a thoroughly good tanker with years of service to come was still a good buy at about £12 million each but it was the failure to control costs that caused the uproar. It certainly did not enhance my career prospects, but I was old enough and ugly enough to take it on the chin. Organizations like the Filton factory were focused more on new aircraft building and had some difficulty in changing direction and refurbishing old aircraft. One aspect of taking on this contract was the re-validation of the Flight Test Organization, for which I still had responsibility, by A & AEE Boscombe Down Boscombe Down.

Roy Radford had succeeded me as Chief Test Pilot and I had recruited John Lewis, Chief Test Pilot of Rolls-Royce to come and join our team. The Technical Department under Mike Bailey had been reduced considerably but still contained several top quality flight test engineers. John Liddiard, ex-BA came in as Flight Engineer, having been on the VC10 in BA, before he joined the BA Concorde Fleet.

There was no difficulty in obtaining A & AEE approval and I believe that my own experience in Valiant-to-Valiant refuelling helped. The first flight of the KMk2 ZA141 was made by Roy Radford on 22 June 1982. The flight trials presented some problems: the Mk32 pod was fairly new and had not done a great deal of flying and trouble with seals, wind in arrangements all came to the fore, but there was no shortage of effort by Flight Refuelling to put the problems right.

The biggest problem occurred on Flight 16. The aircraft was preparing to demonstrate rate of roll at the corner point of the Mach no. (MD/Airspeed (VD)) of the flight envelope starting the dive from 38,000 ft with speed brakes extended at 24,000 ft. At the corner point the abrupt change in the slope of the line made it very difficult to follow without temporarily exceeding the limit. On this occasion extension of the speed brakes started a lateral excitation of the tailplane, increasing in amplitude. On selection of the air brakes in, the oscillation damped out but at a speed of 386 k a large nose down trim change occurred requiring a pull force of 135 lb. Recovery was successful but a speed excess of VD plus 24 k occurred at 18,800 ft. This did not surprise me from my own experiences during VC10/Super VC10 dives. Examination after recovery by means of the periscope showed damage to the fibre glass fairing forward of the tailplane and the top engine panel from no. 3 engine was missing. After flight inspection revealed major damage to the tail section including three cowling panels missing from no. 3 engine. But the principal damage was to the fin rear span web and supporting structure in the region of hinge rib no. H4 together with the tailplane bullet structure, which is an integral part of the fin. A contributory factor to the failure of the rear span web was the pre-existence of a fatigue crack approximately 0.8 in long and under annular re-enforcing plates around an inspection hole in the web. The fatigue crack had a significant effect on the strength of the web. Without the crack, the particular loading at the critical point of failure would not have been high enough to cause failure. The crack was concealed under the re-enforcing plates and had not been revealed by normal visual inspection techniques.

The fin was changed from another VC10 so that the programme could continue. A great deal of analysis took place following this incident. The main solution was provided by reducing the margins between the maximum normal limits VC/Mc and VD/Md. The original margin had been applied to the VC10 some years ago under an arbitrary fixed value. Later revisions of the Air Worthiness Rules reduced this margin. Consequently Vd/Md for the VC10 was reduced immediately to comply with up-to-date margins, and was a case of good common sense prevailing at long last. One person who should have known better tried to blame the exceeding of the speed limits as the primary cause. This was not justified and the full investigation disproved this theory in every respect.

The RAF formed No. 101 Squadron to receive the tankers as they came into service and this operation continues today. I made a few flights with Roy Radford and John Lewis and my last flight in a Super VC10 took place on 15 September 1985. After the flight John Lewis remarked, 'I see you haven't lost your old touch!' – a happy note on which to finish what at times had seemed a long journey.

TEN

The British Aircraft Corporation and the 1-11

Looking back it is remarkable that BAC ever survived the mid to late 1960s. The fact that it did was largely due to a very brave decision in May 1961 to go ahead with a new jet liner, the BAC1-11, on the back of the initial order for ten from Freddie Laker of British United Airways.

At the time of the formation of BAC the Weybridge design effort was concentrated on a medium-range, scaled-down version of the VC10 with four RB163 engines. The VC11 was originally an entrant in the BEA competition which went to the Trident. Government launch aid had been procured for the project; TCA had signed a letter of intent for fourteen aircraft; and Sir George Edwards had high hopes of doing business with his old friend, TCA President, Gordon MacGregor.

The VC11 was a direct competitor to the Boeing B-727 and it was clear that competition would be very fierce. I remember taking part in some of the early meetings on the project and regarded some of the take-off and landing speeds on the VC11 as being rather high. Lengthy decision making turned some potential customers away to the Boeing B-727. Nevertheless, the VC11, although too big for BEA, was about the right size for the world market.

As a result of the merger of Hunting into the British Aircraft Corporation, a small twin-engined project carrying about fifty passengers using Bristol Siddeley BS 75 engines of

7,000 lb each, known as the Hunting 107 appeared on the scene. A lot of internal debate on the VC11 versus the Hunting 107 followed, as it was not possible to continue with both projects. Although the VC11 had two points in its favour, firstly the Government had agreed launching aid and secondly it had TCA as a customer, support for a twin-engined jet was gathering impetus within BAC.

The result of an extensive world market survey covering eighty-nine, then extended to a hundred, airlines determined that the H107 could be very successful if it was made bigger and equipped with more powerful engines. And so the H107 was re-designed into the BAC 107 and then BAC1-11 powered by two Rolls-Royce RB163 engines, which became the Spey. The choice of engine was rather a blow to Bristol Siddeley Engines, 50 per cent owned by Bristol Aeroplane, a BAC parent. The 'little' aeroplane, Hunting/BAC 107, remained in existence for a couple of years or so but nobody bought it.

The BAC Board decision to go ahead with the BAC1-11 was made in May 1961 and it put down a first batch of twenty aircraft while formally dropping the VC11. This decision was not easily reached because of Vanguard and Britannia experience as well as the behind the scenes rumours about the VC10. The BAC1-11 became the mainstay of BAC following the cancellation to TSR 2 in May 1965.

The significance of Freddie Laker's order for ten with five options cannot be over-emphasized. This was a private venture product ordered by a private venture independent airline, albeit with 50 per cent launching aid promised, which kept the initial risk within the bounds of BAC contemplation. The BAC1-11 was designed from the start for the world market covering the heavily travelled short-haul routes of the world. The object was to bring jet comfort and speeds to routes served by propeller-driven aircraft.

The first model of the BAC1-11 was the 200 Series, carrying seventy-nine passengers and with a take-off weight of 78,500 lb powered by two Rolls-Royce Spey 506 engines.

The BAC1-11 used the experience of the Vanguard and VC10 in utilizing 'milled from the solid' or 'sculptured' components in place of fabricated parts built up by riveting or bonding. The use of machined skins, milled from solid alloy planks was introduced on Vanguard wings and then on the VC10. This advanced process was used for many fuselage components, providing increased fatigue life and ease of inspection. The split system philosophy of the VC10 was adopted whereby the aircraft working systems were split into continuously operating halves, either of which could keep the aircraft operating normally without loss of performance. Fail-safe engineering practices were used throughout. Special attention was paid to quick turn round capability, high utilization, low costs and low break-even loads.

The person chosen to head the BAC1-11 sales effort was Geoffrey Knight, assisted by Derek Lambert. Geoffrey Knight had sold Britannias to Freddie Laker and was in the process of selling him VC10s. Engineering came under Basil Stephenson, Director of Engineering looking after the design teams at Weybridge and Hurn with Fred Pollicutt, Technical Director covering Luton's team headed by Ken Carline, Chief Designer. Arthur Summers, formerly Managing Director of Hunting, had overall responsibility for production.

Geoffrey Knight endured some hard bargaining with Freddie Laker concerning price after the specification was agreed. In fact Geoffrey Knight recalled the 'horse trading' that went on including issues like pilot training, engineer training costs and publicity support. The deal was finally settled at approximately £709,000 per aircraft between races at Sandown Park. I have always found this a most incredible figure when compared to today's prices, which have become so high.

At the start of the 1-11 there was a gap in American coverage for the type of operation envisaged. The sale to BUA was quickly followed by an order from Braniff, starting with six and rising to fourteen. Braniff was followed by

Mohawk for four aircraft and we saw a great deal of Mohawk's President, Robert Peach. Eventually Mohawk operated twenty 1-11s.

While the 200 Series had commenced final assembly at Hurn, earlier interest from American Airlines came to the fore. The advent of the Spey 511 engine of 11,400 lb thrust instead of the 10,410 lb of the Spey 506 enabled BAC to offer the 300/400 Series with a capacity of 85 pax and with take-off weight increased to 92,000 lb and an increase in range. This aircraft was very suitable for American Airlines – one of the big four US carriers. Geoffrey Knight and Derek Lambert practically lived in New York, while fairly frequent visits to the President, the legendary airline figure of many years standing, C.R. Smith, were made by his old friend Sir George. C.R. Smith was a long-time member and founder of American Airlines. Eventually breakthrough occurred on 17 July 1963 when American ordered fifteen 400 Series, later increasing the order to twenty and then thirty.

One factor that undoubtedly helped the US sales was the existence of a considerable sales, servicing and spares organization at Arlington (Virginia) on the edge of Washington National airport under the leadership of Murray White. This facility, which had been started for the Viscount, offered an 'over the counter' service of airframe spares from a nut and bolt to a whole wing. Engine support was provided by Rolls-Royce Canada. In later years BAC (US) expanded and moved to a new impressive facility at Dulles airport with many key players including Bob Gladwell, Bernard Brown, Dewi Rowlands and Brian Thomas. During the years 1977–8 the company became British Aerospace Inc.

On 20 August 1963, one month after the American order, the prototype 1-11 GASHG made its maiden flight from Hurn in the hands of Jock Bryce accompanied by my deputy, Mike Lithgow. By this time the 1-11 had been ordered by BUA, Braniff, Mohawk, American, Central

Africa Airways (two), Kuwait (three) and Aer Lingus (four) – a total of fifty aircraft. After a couple of flights at Hurn, GASHG came to Wisley.

I had my first experience of the aircraft on 2 September 1963. I remember very well that I did not like the feel of the elevator control which was a spring tab arrangement and I thought that there was a distinct lag between control input and aircraft response. My comment was not very well received but several other test pilots, notably John Cochrane and Roy Radford, thought the same.

The flight test programme went along at a tremendous pace and fifty-two flights covering eighty-one hours flying had been completed by 22 October 1963. Then disaster struck. GASHG took off from Wisley with test pilots Mike Lithgow and Dickie Rymer with a test crew of five, Dick Wright as Senior Flight Test Observer, Gordon Poulter, Flight Test Observer, D.J. Clark, Flight Test Observer, B.J. Prior, Assistant Chief Aerodynamicist and C.J. Webb, Assistant Chief Designer for stalling tests.

I was sitting in my office at Wisley when ATC rang through to say they had heard from Boscombe Down that the 1-11 had crashed at Gratt Hill, near Cricklade, Wiltshire, quite near to Boscombe Down. My first thought was to get hold of Jock, who had been called to a special Board meeting. I telephoned to Weybridge to be told that he was engaged at a Board meeting. I recall snapping down the telephone at an innocent secretary 'Just get him out! This is an emergency!' Jock came over to Wisley immediately so that we could decide on our plan of action.

Further details of the accident came through very quickly. There was clearly no point in flying to Boscombe Down, that came later. Our first job was to deal with the families, as we knew by then that there were no survivors, and Mike Crisp, the Flight Test Manager offered to help. Between us we visited all the wives and broke the news. I went to Dorie Lithgow who at first was not prepared to believe me. When I arrived she was talking to a girlfriend on the telephone, so I picked up the telephone and said to this person 'Please could you come here straight away', which she duly did. By evening we had done all that we could.

The scene of the crash was not a very pleasant sight as it had burnt after impact. A busybody became involved and threw some doubts as to where each person was sitting and so it became necessary to obtain dental records. This simply caused further suffering to the widows and their families. There were more than twenty children affected by the tragedy but BAC formed a trust to continue their education as it had been originally planned.

The circumstances of the accident were fundamental to the stalling of T-tail aircraft. The conclusions of the Ministry of Aviation's Official Accident Report read:

i.   The aircraft was flying in accordance with the B Conditions of the Air Navigation Board 1960, it had been certified as safe for the flight and was properly loaded.

ii.   The pilots were appropriately licensed and were experienced in experimental flight test work.

iii.   There was no evidence of any pre-crash structural failure.

iv.   The nose down pitching moment (elevator neutral) just beyond the stall was insufficient to rotate the aircraft at the rate required to counteract the increase of incidence due to the g-break.

v.   During the fifth stall, the angle of incidence reached a value at which the elevator effectiveness was insufficient to effect recovery.

Opinion During a stalling test, the aircraft entered a stable stalled condition. Recovery from which was impossible.

It was no good pretending that we had not had an accident. It had been seen by all the world. It seemed to me that one of the main things to do was prevent anyone else making a similar mistake. After talking to Sir George Edwards and Jock Bryce, it was agreed that Ken Lawson, Chief Aerodynamicist and myself should go to the United States armed with the accident traces and other data to talk to Boeing, Douglas, Lockheed and Lear. We started with Boeing Flight Test under Dix Loesch. Having shown them all our results, they asked me what I would like to do. I saw the B-727 Project Pilot, Lew Wallick, go white when I said I would like to do some stalls on the B-727. I explained that I had not come to be a hero and that I would follow the same procedure as I was using on the VC10 and would stop at whatever angle of attack that I was given. This relieved the tension and indeed I made two flights with Lew in a B-727. I learnt that the B-727 had entered a stable stall on one occasion but due to the hydraulically powered elevator, recovery had been effected, albeit with a height loss of nearly 10,000 ft. For good measure, I had a couple of flights in a B-707 as well. The B-727 experience convinced me that our ARB would call for a stick pusher and indeed this is what did happen when the B-727 came on to the British register. Subsequently this requirement was not demanded on later B-727s.

George Jansen, Chief Engineering Pilot of Douglas Aircraft who were developing the DC9 came up from Long Beach and went through the 1-11 accident record. As a result, I believe that it is true to say that Douglas decided to put a bigger tailplane on the DC9. Similar meetings took place with Lockheed and Lear. We could not stay too long, as it was necessary to get back to the VC10 stalling programme, which as previously mentioned nearly ended in tears on 31 December 1963.

In the meantime, Hurn was nearly ready to start producing production aircraft. In the light of GASHG's accident, BAC had to revamp the 1-11 programme and to determine the action necessary to stop the same thing happening again. A further difficulty occurred when Jock Bryce failed his annual medical, an event which most of us dreaded as so much was at stake. This of course meant that all 1-11 flying experience had virtually disappeared except for my own rather limited exposure. I had already recognized that we needed another pair of well-proven hands, so I turned to Peter Baker, with whom I had done some Valiant flying at Boscombe Down and who was currently working for Handley Page. He agreed to join us at Wisley and in due course I made him 1-11 Project Pilot, knowing full well that I would have to lead the stalling programme when it recommenced.

Following the accident to GASHG an intensive programme of development work was carried out, resulting in a number of configuration changes, which included:

i.   A chordwise extension of 2½ per cent to the wing leading edge along the entire span of modified profile.

ii. A redesigned and repositioned leading edge fence.

iii. Fully powered elevator control replacing the original servo tab manual system. Initially this was not fully duplicated and a hydraulic system failure resulted in one elevator being powered while the other went into manual. I thought this was a half-baked solution and said so. In due course a fully duplicated system was provided and only after a double hydraulic failure did the elevators revert to manual.

iv. An electro-pneumatic stick pusher system was fitted.

The stalling trials resumed in August 1964 on the fourth BUA aircraft GASJD. Peter Baker and I flew a number of flights together, completing a further thirty-nine stalls with satisfactory results. Sir George knew that I had a holiday planned in Lairg, Sutherland with Colonel and Mrs Bibby, who owned a lovely estate overlooking Lock Shin. He insisted that I needed a bit of rest and felt that the programme could well be left in the competent hands of Peter Baker. I was not sure about taking leave at that particular time, but I went after some persuasion.

I was fishing during the afternoon of 28 August 1964 when Mrs Bibby suddenly appeared and told me to telephone Ollie Oliver at Wisley immediately. Ollie had come from Hunting and by that time had become my deputy. I went back to Lairg Lodge and spoke to Ollie who said 'I am afraid that I have some bad news, GASHD is in a field north of Boscombe Down, nobody is injured but there is an aircraft on its way to RAF Lossiemouth to fetch you'.

Lairg to Lossiemouth is about 100 miles and not an easy journey, but I managed to find a local taxi who was prepared to take me. I thought the driver was somewhat erratic on the way to Lossiemouth and subsequently learnt that he committed suicide on the return journey. It took about four hours to get there where one of the Heron communication aircraft was waiting. Without further ado we left for Wisley, arriving at about 0400 hrs. When I arrived, I found several members of the Aerodynamics Department and Flight Test personnel poring over the flight traces with very puzzled faces. They could not see anything wrong with the aircraft.

I stayed at Wisley until Sir George and Charlie Haughton, now Managing Director, appeared in my office. We were well into a somewhat lengthy discussion when the door suddenly opened and Peter Baker walked in. He came straight to me and said, 'I am the biggest bloody fool in the business, I have slept on it and there was nothing wrong with the aircraft'. I responded by telling him to go to his office where I would join him in few minutes. After Peter left I turned to Sir

George and Charlie and said, 'I want to keep him'. They agreed to this and Peter Baker stayed.

My decision was criticized for being too soft, especially by some of the sales department. I made my decision for a number of reasons. Firstly, good test pilots are hard to come by and Peter was well within the scope. Secondly, I did not believe that one mistake made under conditions of high risk and high stress should automatically constitute dismissal and thirdly, more help should have been provided by the other pilot, which made me conclude that the crew could have been better constituted. Geoffrey Knight was not in agreement with me and in due course when major sales tours were planned, we agreed to replace Peter with Roy Radford. Peter would eventually become a part of the Concorde team. Years later, when the Concorde programme was coming to an end, Peter became Chief Test Pilot of ARB for several very successful years. His intimate knowledge of rules and regulations were of great value to them.

The official findings of the accident investigation were:
i. The aircraft was flying in accordance with the B Conditions of the Air Navigation Order 1960: it had been certified as safe for the flight and was properly loaded.
ii. The pilot and co-pilot (T.S. Harris) were properly licensed and were experienced in experimental flight test work.
iii. No evidence of pre-crash malfunction or defect was found in the aircraft.
iv. When the pilot pushed the control column forward after the stalling run (clean, forward centre of gravity), the aircraft responded normally, resulting in a marked reduction of normal acceleration (g).
v. Although the aircraft's behaviour and instrument information indicated otherwise the pilot believed

the aircraft to be developing a stable stall condition and streamed the tail parachute.

vi. The nose down pitch due to the tail parachute was small because the angle of attack (incidence) was low.

vii. Had the tail parachute been jettisoned during the descent, the flight could have been continued normally.

Opinion: During a stalling test the pilot streamed the tail parachute under an erroneous impression that the aircraft was in a stable stall: an emergency landing was necessitated by the retention of the tail parachute.

This accident, which was no fault of the aircraft itself, was regarded in the outside world as '. . . another 1-11 stalling accident' and there was considerable coverage of it in the UK and US media. This did a great deal of harm to the 1-11 sales prospects. When one couples this accident with a third mishap on 18 March 1964 when GASHJB flown by Staff Harris and Dinty Moores made a very heavy landing at Wisley, causing considerable damage, it is amazing that the project managed to survive.

The modifications following the first accident had to be fitted retrospectively to the first three production aircraft, which flew with the original elevator arrangement. Just before touch down, Staff Harris got out of phase with the slightly peculiar elevator control and this resulted in a pilot induced oscillation P10 developing. It was reported that intervention by the other pilot made the situation worse. I was in the Aircrew Mess having lunch when I heard the crash siren sound. I rushed out on to the apron and looking towards the runway I could see a sad looking 1-11. One side of the tailplane practically broken off and one engine was sitting on the runway. When I got up to the runway Staff Harris was walking round the wreck saying, 'I just don't know what happened'. Without thinking, I responded very unsympathetically by saying, 'It bloody well looks like it'. Attempts to rebuild GASJB were eventually aborted.

There is no doubt that several sales, especially to Alitalia, which was nearly settled, floundered on the back of these accidents. The situation really was very bleak, but nobody gave in. BAC teams were sent to all customers and potential customers, taking with them the flight records and as they were talking to professionals, it was soon possible to prove what had happened and gradually confidence in the 1-11 was restored. BAC's competitors, especially Douglas, made the most of the market advantage that they had with the DC9, even thought it was also T-tailed.

While GASJA carried out performance trials, the third 1-11 GASJC resumed the stalling programme in September and was used for most of the critical certification tests. The fitting of artificial stall warning and of stick pushers sounds more straightforward than it actually is. On the one hand there is the need to meet the regulations and afford the necessary protection, while on the other there is the need to make sure that the devices do not give nuisance warnings and thus restrict optimum speeds for initial climb and approach. The importance of minute speed changes on the approach had for instance an effect on maximum permitted landing weight at a particular airfield. It was necessary to ring out the last drop, so to speak, on the 1-11 as competition was so fierce. Roy Radford, who had taken over as BAC1-11 Project Pilot, did a wonderful job in this respect, becoming Assistant Chief Test Pilot on the 1-11 later on before joining in on the Concorde programme. In 1982, he would succeed me as Chief Test Pilot.

Tuning of the 1-11 stall protection system went on for years along with aerodynamic refinements. Many attempts were made to fox the stall protection system by ARB and FAA test pilots together with one of the American Airlines development pilots, Captain Glen Brink, who really upset his boss with his antics. Dave Davies, Chief Test Pilot of the ARB had become renowned for his tough approach to certification, which prompted his counterpart in the FAA, Dick Slief, to really frighten me on one occasion. We were carrying out a single-engined climb at take-off safety speed (V2) when he suddenly pulled the stick back as hard as he could. The aircraft shook, the pusher fired and the engine fuel dip system, which had been incorporated to deal with engine over-temperaturing in stalls at high engine power, operated. In other words, the system worked perfectly. Dick Slief looked at me and said 'Is Dave Davies any tougher than me?' I had known Dick for a long time, dating back to my visit to Boeing, but it did not stop me thinking 'You stupid bugger!'

Another particular incident which I remember was when Dave Davies was carrying out simulated engine failures during take-off. On this occasion he asked me to stop cock the right-hand engine at quite a low speed during the take-off run. The test was absolutely satisfactory but when it came to re-starting the failed engine, its speed had run down to almost zero and it was necessary to accelerate to 300 k before the wind-milling speed was sufficient to permit an air re-start. Approximately 1,000 stalls were carried out on the 1-11 200 Series to satisfy the ARB and FAA and ourselves. Other production aircraft, notably GASJI (used for route proving) GASJC, GASJF and N1543 (Braniff) were stalled to confirm that production variability had not caused significant differences in handling characteristics in the stall.

The scope of the tests carried out were:

i. Slow straight stalls – speed being reduced at 1 k/sec.

ii. Slow turning stalls – left and right – 30° bank applied and speed reduced steadily at about 1 k/sec.

iii. Slow dynamic stall – bank angle about 40° – with normal acceleration increased up to about 1.5 g at the stall.

iv. Fast dynamic stall – bank angle about 40° – with the application of normal acceleration delayed until approximately the stick shaker speed, when a rapid application of elevator was applied causing a rapid build up of angle of attack and normal acceleration at the stall. Stalls on one engine were also carried out and these presented no difficulties from the aircraft handling point of view.

Engine handling in the stalls using flight idle power was entirely satisfactory although minor fluctuations of engine revs per minute (RPM) were noted at the 'g' break angle of attack. With engine power on, there was a tendency for the engines to surge and cause a sharp drop in RPM and an unacceptable temperature rise necessitating throttle lever action to recover the engines. A fuel dip system was developed to overcome this problem.

The stall protection system consisted of three main elements:

i.   Stall warning – the firing point being advanced by rate of change of angle of attack.
ii. Auto-ignition – automatic switch on of engine igniters at an angle of attack below the stalling angle.
iii. Stall identification – known as the stick pusher – providing positive identification of the stall and to initiate recovery in order to avoid attaining high angles of attack. All configurations of flap, airbrake, power, weight and altitude were covered.

When I add this programme to my Vanguard and VC10 experiences, it was not difficult for me to feel fed up with stalling aircraft. So I decided to give a lecture to the Royal Aeronautical Society on the subject.

We now had to turn our attention to the 400 Series,

the first of which was GASYD, a company owned aircraft, which remained with BAC and then BAe for many years into the 1990s.

Several of the 200 Series used in the development programme went back to Brooklands for refurbishment. The return of both VC10s and BAC1-11s into Brooklands became a regular event. Certification of the 200 Series by the ARB was granted on 9 April 1965 while the FAA type certificate was given on 16 April 1965. This enabled BUA to commence the first revenue flight on a 1-11 on 9 April 1965, having received delivery of the aircraft three days earlier flying from Gatwick to Genoa while Braniff flew their inaugural service on a multi-sector flight between Corpus Christi and Minneapolis-St Paul. All fourteen of Braniff's order, all ten of the BUA order, the four for Aer Lingus and the first five for Mohawk Airlines were all delivered in 1965.

Two 400 Series aircraft were built ahead of the American Airlines order. GASYD flew on 13 July 1965 and GASYE followed in September 1965. The first aircraft was used for tropical trials at Madrid-Torrejon while the second was prepared for an intensive programme of demonstration flights through the world, incorporating an executive layout in the front and airline seating in the rear. The first tour started from Wisley on 17 November 1965 for Central America and the USA, visiting thirty-three central cities in the USA, Nassau in the Bahamas, Mexico City, Guatemala City, San Salvador, Tegucigalpa, Managua and San José in Costa Rica. A three-week break in the tour was made to permit American Airlines to do some crew training. The FAA type certificate for the 400 Series was granted a few days after the tour started on 22 November 1965. The aircraft returned to the UK on 8 January 1966, having flown 50,000 miles. It was off again on 21 January 1966 bound for Rome, Damascus, Bahrain, India, Bangkok, Singapore, Bali, Darwin, Townsville, Sydney and finally Christchurch on 26 January 1996. Demonstrations were also made in Australia, Japan, Manila, Hong Kong, Rangoon, Colombo and Tehran. It returned to the UK on 8 March 1966 having flown 70,000 miles but not for long as it was soon off again visiting South America and USA: covering Nassau, Freeport, Miami, Port of Spain; also Belem, Brasilia, Rio de Janeiro, São Paulo, Urububunga, Landrina, Porto Alegre and Curitiba, all in Brazil. Then on to Paruguay, Uruguay, Buenos Aires, Santiago and Lima in Peru, Venezuela, Ecuador. Another 40,000 miles were completed.

In 334 flights on the three tours GASYE flew 160,000 miles under all temperature conditions from arctic to tropical, visiting some airfields as high as 9,000 ft above sea level. Over a hundred guest pilots flew the aircraft and a total of 396 hours had been added to GASYE's flying hours. This was a clear demonstration of the rugged and reliable nature of the 1-11.

The whole of the 1-11 programme in its inception, design, sales, production, after-sales services and customer training schemes was something to be proud of mainly because of Geoffrey Knight's knowledge of what was needed. It was as good as anything offered by the American manufacturers. The pilot strength of Wisley and Hurn peaked at thirty-five pilots, most of whom were training captains in order to cope with the size of the programme. During 1966, the 400 Series entered service with American Airlines enabling the inaugural service to commence on 6 March 1965. By the end of the year all thirty aircraft had been received.

On 4 November 1965 I flew Prince Philip from London airport (Heathrow) to Toulouse with Dennis Hayley-Bell, our senior BAC1-11 training captain as co-pilot. This was my first visit to Toulouse since my assignment to the Concorde programme and a splendid occasion it was. There was a special car for the pilots supported by two outriders on motorcycles, followed by a wonderful dinner at the Sud Aviation (now Aerospatiale) guest house. I may say that my next trip involved taking a taxi from the airport side to the

Sud factory on the other side of the airport. Prince Philip flew the 1-11 back to Hurn and then we dropped him off at Gatwick before returning to Wisley. BAC/BAe tried to sell 1-11s to the Queen's Flight on several occasions but it never came to anything.

BEA were now beginning to show an interest in the 1-11, so I took Captain A.S. Johnson, who had been Flight Manager of the Vanguard fleet, to assess the suitability of the 1-11 for operation in to Jersey in July 1966 in GASYD. This had been preceded by a flight from Heathrow to Hurn and back, for the BEA Board, earlier in the year. We carried out a series of take-offs and landings without any problems and BAC hoped that BEA would accept the aircraft off the shelf but this proved to be wishful thinking. In fact the changes required by BEA were the equivalent in cost terms to the price of another aeroplane. All the switches in the cockpit and elsewhere had to be changed round to match their other types. I for one was thankful that the 1-11 had got started without them.

By this time I was heavily committed to Concorde and all the preparations for the first flight. Consequently, my contribution to the 1-11 became very limited although I did make the first flight of the 500 Series on 8 June 1967 from Hurn. Geoffrey Knight was most insistent on my doing this, although I felt that it should have been left to Roy Radford. Anyway we did the early flights together before I more or less faded out of the 1-11 picture in terms of flying but not responsibility.

The prototype 500 was in fact GASYD with its fuselage lengthened by 13 ft 6 in and a 5 ft increase in span. Take-off weight was increased to 92,453 lb and later 104,500 lb thanks to the Spey 512 engine providing over 1,500 lb thrust increase followed by a further 550 lb of thrust. A number of aerodynamic refinements were incorporated and there was an all-round improvement in mission performance. The original concept for the 500 was to meet BEA's German and domestic routes, but thanks to the extra thrust (550 lb) a capability of meeting the rapidly expanding inclusive holiday market requirement of 1,000 to 1,500 mile radius from the airport of origin was achieved, carrying 119 passengers.

Unfortunately, that was the end of Spey development which prevented a further stretch of the 1-11 up to 130 seats or more. Thus the 1-11 could not respond to the challenge of the four stretches of the DC9 with the JT8 engine up to 137 seats and eventually 160. This prompted BAC to move towards a new version of the 1-11 known as the 475 Series, aimed at bringing jet travel to secondary airfields where the runways were either unmetalled or gravel requiring short take-offs and landings and above average descent and climb-away paths. GASYD was again used as the prototype with its fuselage returned to 400 Series standard but with the 500 Series wing and the 12,550 lb static thrust Spey 512-14DW (designated wet) engine.

The underneath of the aircraft and the flaps were protected against flying gravel and deflectors were fitted to the nose and main landing gear. A large cargo door was also incorporated. Water Beach airfield was used for the rough runway tests. This aircraft gave some fabulous demonstrations on all sorts of unprepared runways in South America, Central America, India and Japan, in the expert hands of Roy Radford, who really did a magnificent job displaying the capabilities of the aircraft. The 475 did not prove to be a commercial success and was only bought by Air Pacific and Air Malawi.

In 1968 Tarom, the Romanian National Airline and the Romanian Government did a deal with BAC whereby they bought six 1-11 400s while they negotiated a joint arrangement with BAC. They also did a deal with Britten-Norman for the local assembly and eventual build of 212 Britten-Norman Islander light transports. This was followed by an order for five 1-11 500s in 1975, which included certain 1-11 and Rolls-Royce engine work being done in

Romania plus another batch of Islanders. At the time of nationalisation of the British Aircraft Industry in 1977, Romania was discussing a deal with BAC to build over eighty 1-11s under licence with the initial batch being built in the UK and eventual construction for the remainder in Romania. Most of these were to be 475s. Subsequently the contract was for a mixture of 500s and 475s. The programme known as Rombac was under the personal direction of the then Managing Director of the Commercial Aircraft Division, John Ferguson-Smith, and the flying aspects were master-minded by Dave Glaser, the Senior Production Test Pilot at Hurn.

The Series 475 development aircraft went to Japan in October 1976 to prove its suitability for operations at the more restricted Japanese airfields. BAC was hopeful of selling this version to TOA Domestic and All Nippon Airways. Many of the airfields had only 4,000 ft runways but the aircraft coped with these conditions to British take-off and landing criteria. The Japanese authorities insisted on a greater than normal fuel reserve and greater performance margins if the type was to be certified in Japan. In order to cope with these requirements a 475D was proposed, which incorporated a further wing tip extension, raising the span to 96 ft 10 in. An extension of the trailing edge flap chord by 4.65 per cent and further re-profiling of the wing leading edge was also included.

Further wind tunnel tests showed that rather than re-design the whole wing, it was possible to achieve the required performance by fitting a small triangular fillet to the area of wing root to the wing fence. Further requirements were proposed, including the use of Hytrol MkIIIA anti-skid system for the brakes, together with automatic braking and deployment of lift dumpers triggered by a combination of the compression of the main oleos on touch down and coupled with wheel spin up. A further improvement in silencing of the exhaust area of the engines was proposed, which incorporated an eight-lobe exhaust nozzle and an ejector cowl which moved aft behind the reverser cascades for take-off and landing.

These changes, with the exception of the ejector cowl modification, were built into GASYD in April 1977 and the aircraft was then re-designated Series 670. A comprehensive flight trial programme was flown on the Series 670. Unfortunately, there was no interest in Japan or any other country and none were built.

GASYD continued to fly in both development and communications roles from Filton to Toulouse supporting the Airbus programme and also for BAe Warton on a regular Munich run. Its last development task entailed the flight testing of a Lucas Fly-By Light spoiler actuation system. The aircraft was retired in October 1993 and was flown into Brooklands on 14 July 1994 to join examples of its stablemates, an early Vickers Viking, a Viscount 800, a Varsity and a standard VC10 at the Brooklands Museum.

Other developments of 1-11 were considered in the mid and late 1970s. The 1-11 600 Series, seating up to 130 passengers and powered by Rolls-Royce Spey giving 17,000–18,000 lb thrust, was offered to British Airways in competition with the Boeing 737/200. BAe proposed that the 600 Series could be re-engined later with Rolls-Royce RB432 turbo-fans in the 16,000 to 18,000 static thrust range under the designation Series 700. Basically the 600/700 was a 500 Series with the revised wing of the Series 670 and engine silencers.

A more ambitions development was the 800 Series which would have been powered by two SNECMA/GE CFM56 engines of 22,000 lb static thrust and fuselage stretched by 32 ft 6 in over the 500 Series providing accommodation for 144–61 passengers. Span would have been increased by 10 ft and there would have been a new centre section.

Both the 700 and 800 were succeeded in 1976 by the proposed X-11 which was bigger, quieter and more fuel-efficient than the 800 accommodating 130–60 passengers. The engines would have been either CFM56 or Pratt and Whitney JT10D providing 22,000–24,500 lb thrust. Weight increased to 140,000 lb but operation from 5,000–7,000 ft runways was possible. Ultimately BAe decided to discontinue the X-11 in favour of a 160-seat airliner known as Jet, which was a collaborative programme between Aerospatiale, BAe, Fokker VFW and Messerschmitt-Bolkow-Blohm. This also failed to emerge.

Nationalization of BAC into BAe did little to develop the 1-11 because of the obvious clash with the HS146. Proposals to fit the Rolls-Royce Tay engine developed from the Spey were agreed between Rolls-Royce and Dee Howard to the extent that one BAC1-11 400 Series retro-fitted with two Tay engines and one of BAC's ex-Concorde Test Pilots, Johnnie Walker, was employed by Dee Howard to conduct the flight trials. The project was abandoned when 90 per cent of the flying trials had been completed.

Even more excitement was generated when a company called Associated Aerospace with whom I had a connection, before it went bust, announced that it was ordering fifty 1-11 500s from Romania powered by Rolls-Royce Tay 640 engines. It was also announced in February 1993 that Kiwi International Airlines of Newark, New Jersey were ordering eleven 500 Series from Romania (Romaero) with an option on a further five. None of this actually happened and although BAe took part in all the discussions, it only gave limited support to the programme. The 500 Series 1-11 with Tay engines was a better aircraft that the BAe146 in the opinion of many but politics won the day. Other versions in the form of the BAC2-11 followed by the 3-11 were seriously considered but for various reasons, they never happened.

After the last Series 475 left Hurn British Aerospace closed the Hurn factory and, soon afterwards, Weybridge, having shut down Wisley sometime before. This marked the end of an

The author standing by Concorde 002 at Fairford, 1969.

The author's mother Queenie and sister Brenda, c. 1926.

An early picture of the author and his sister Brenda at home, Caerdelyn, Llanelli, c. 1930.

The author's father Harold in Royal Engineers uniform at the end of the First World War.

Harvard AT6 used for advanced training at No. 4 British Flying Training School, Mesa, Arizona, 1943.

Fellow cadets celebrating at the Jude, Hollywood, 1943. Left to right, Colin Meikle, the author, Desmond Muirhead, Peter Parker.

The author at RAF Abingdon on arrival from South Africa with Rhodesian Ridgebacks Banshee (right) and Hoolie, 1947.

Vickers Viking pictured in front of Table Mountain during the Royal Tour, 1947.

Air Vice-Marshal Sir Edward Fielden, Captain of the King's Flight and Wing Commander E.W. Tacon, Commanding Officer, RAF Benson, 1948.

The one and only Nene Viking flying at Wisley, 1950.

The team of test pilots at Vickers-Armstrongs, 1950. Back row, left to right, Bill Fell (Navigator), Jock Bryce, Mike Lithgow, Guy Morgan. Front row, the author, Les Colquhoun, Mutt Summers, George Lowdell, David Morgan.

Viscount prototype fitted with two Rolls-Royce Tay engines, which the author flew at Farnborough Airshow in 1950.

The 'Black Bomber', Valiant Mk 2, of which only one was built. It was flown at Farnborough at 500 k in September 1953.

Valiant to Valiant refuelling, 1957. The author was flying the receiver WZ390. Note proximity of receiver to tanker.

Avro Vulcan, one of which was used as an engine test bed for the Concorde Olympus engine, 1968.

The first Vickers Vanguard prototype flown by Jock Bryce and the author from Brooklands to Wisley on 20 January 1959.

The author landing a Vickers Viscount 810 at the Farnborough Air Show, 1958.

The Handley Page 115 Low Speed Research aircraft located at Aero Flight, RAE Bedford, 1967.

BAC1-11 500 Series prototype flown by the author and Roy Radford at Hurn, 1967.

Vickers VC10 first production aircraft in BOAC livery, maiden flight 8 November 1962 flown by the author from Brooklands to Wisley.

BAC 221 modified Fairey Delta carrying Concorde ogee wing flying from Filton, 1968.

Convair B-58 Hustler on which the author had ten hours familiarization flying up to Mach 2 at Edwards Air Force Base, California, 1968.

The author and crew boarding Concorde 002 for a test flight wearing full protective clothing. Left to right, the author, John Cochrane (Co-pilot), Brian Watts (Flight Engineer), Roy Lockhart (Navigator), John Allan (Senior Flight Test Observer), Mike Addley, Peter Holding (Flight Test Observers).

First flight of Concorde 002 from Filton flown by the author, 9 April 1969.

The author, André Turcat, Sir George Edwards after a flight in Concorde 002, 1969. Sir George was the first person who was not a member of the Flight Test Organization to fly in the aircraft.

HRH Prince Philip and the author on the flight deck of Concorde at Fairford, January 1972.

The first three Concordes photographed at Fairford, 1972. 01 was the first pre-production aircraft between 001 and 002. Note the difference in the nose/visor.

202 G-BBDG in flight for the first time, flown by the author from Filton to Fairford, 13 February 1974.

The author, Sir George Edwards and the author's wife Yvonne at Weybridge before Sir George retired in 1975.

The author with his two Shetland ponies Caspar and Senator at home, 1996.

The dreaded pair. The author's Springer Spaniels, brothers Charlie and Clayton, 1996.

era. The total UK production of the 1-11 amounted to 235 aircraft of which 232 were delivered to customers. These comprised 46 200s, 9 300s (UK version of the 400), 69 400s, 12 475s and 86 500s. The question will always remain – why was more not done to develop the 1-11? The answer lies in the inability to find a suitable engine, no doubt exacerbated by the bankruptcy of Rolls-Royce in 1970 and the lack of drive for its future when Geoffrey Knight and John Ferguson-Smith were no longer at the helm.

ELEVEN

Concorde: the Background and Early Stages

The peak of my aviation career came with Concorde and I realize my good fortune in being selected for the job by Sir George Edwards. It is a fantastic experience to have had and I look back on over twenty years of intimate involvement with the project with considerable pride and happy memories.

One of the greatest aspects was the human side. The whole team on both sides of the Channel were totally dedicated and were not prepared to be knocked off their perch by the many critics with whom we had to contend. It was an entirely new dimension to work with the French and many lessons emerged and were learnt for the Airbus. I had to learn to deal with the media, some parts of which were friendly, others decidedly hostile and seldom accurate. At the beginning, Concorde probably meant more to the French than it did here but this soon changed when the British public started seeing the beautiful bird, even though the politicians may have had a different view. It is now worth recording how Concorde came about.

The Supersonic Transport Committee under the chairmanship of Morien Morgan (later Sir Morien Morgan), a Deputy Director of the Royal Aircraft Establishment Farnborough, was formed in 1956. The RAE had already carried out sufficient research and design studies to indicate that economic operation of supersonic transports might well be a practical possibility. Membership of the committee included all the major airframe and engine companies, BEA and BOAC, Ministry of Transport and Ministry of Supply. The Air Registration Board, the Aircraft Research Association and the National Physical Laboratory joined later.

Initial interest pointed towards two lines of thinking. One for a medium-range aircraft cruising at about M=1.2 and a long range transatlantic version cruising at M=1.8/2.0. Various configurations were studied. The slender delta figured prominently. Attention was also directed towards structural aspects including the effect of kinetic heating, type of engines, operational flexibility to operate within the current environment, vibration and flutter considerations and commercial viability. The adoption of speeds of about M=2.6 soon lost favour because the committee considered that kinetic heating and the low-speed characteristics anticipated would result in a later entry into service date and higher cost than on an aircraft cruising at M=1.8/2.0 or 2.2.

The committee firmly believed that the use of aluminium alloys would permit this country to be among the leaders in the supersonic transport field. This meant restricting cruise speed to about 1200–1300 mph. Anything faster should be regarded as second generation. The committee addressed the matter of noise and sonic boom. The suggested noise level was the equivalent of about 103 decibels and the sonic boom might well be acceptable above 35,000 ft. Powered flight controls with artificial feel were considered to be essential rather than manual controls because of the movement of the wing's centre of pressure at supersonic speeds. From the work of the committee and its sub-groups emerged recommendations to build two prototypes, one flying at M=1.2 over 1,500 miles and the other flying at M=2.0/2.2 over 3,450 miles.

The French had been conducting similar studies during the 1950s, which came out in favour of the medium-range version, the Super Carvelle SST. Between 1959 and 1963, the STAC changed its view and came down heavily in favour of one type in collaboration with France. Feasibility studies were awarded to Hawker Siddeley and Bristol Aircraft, who were contracted to study collaboration as well as the aircraft itself. Bristol first came up with the Bristol 198 with six engines carrying 130 passengers. The six-engine configuration lost favour on technical grounds and re-emerged as the Bristol 223 with four engines.

When the initial meetings between BAC and Sud started in the early 1960s, the concept of two versions still remained and indeed a meeting between Dr Russell (Sir Archibald Russell) and Dr Bill Strang of BAC with Pierre Satre and Lucien Servanty of Sud confirmed that a common basic aircraft in the two versions was feasible. On this basis an agreement was signed on 29 November 1962 between the British and French Governments to 'develop and produce jointly a civil supersonic transport, on the basis of equal sharing between the two countries'. The two versions were specified and it was assumed that two prototypes and two pre-production would be built. The agreement was signed by Julian Amery (Sir Julian Amery), Minister of Aviation and by the French Ambassador Geoffroy de Courcel and contained a no break clause, although the Treasuries on both sides wanted one.

Julian Amery argued that it was not reasonable for either side to commit such substantial resources for one partner to drop out in the middle of the programme. He suspected a French withdrawal whereas it was the British, following a change of Government in the Autumn of 1964, that tried to cancel. Roy Jenkins (Lord Jenkins), the Minister of Aviation was sent with clear cut instructions to inform the French Government that the British Government wished to cancel the project on the grounds of costs, which had already doubled from the original estimate of £140 million to £280 million.

The French strength lay in the 'no break' clause and they used it. Harold Wilson was warned by his Attorney-General that unilateral withdrawal could result in substantial damages being awarded by the International Court at the Hague and the doors to the Common Market would be closed forever. General de Gaulle, the French President, had big plans for the glory of French civil aviation. So Roy Jenkins came home with a bloody nose and Harold Wilson had to admit defeat. By this time the original idea of a medium-range version and a long-range version had been overtaken after eighteen months of wrangling and had been cancelled in favour of the long-range model. This was itself substantially different from that defined in the Treaty Agreement. The change of definition was the main contribution to the rising cost estimates.

What has not been generally appreciated until the recent publication of Kenneth Owen's book, *Concorde and the Americas*, is the considerable pressure exerted by the US Government on the British Government to cancel the Concorde project. It is difficult to be sure why this attitude was adopted but I believe the simple reason is that some of the senior politicians in the USA did not want supersonic transports at any price and the sooner the idea was put to bed the better. With this in mind, it is not surprising that the elapsed time to obtain permission for the Concorde to land in the US, firstly at Washington Dulles airport and secondly at Kennedy International airport New York was so protracted. The delay in obtaining clearance to use Kennedy had the most profound effect on the commercial success of the Concorde and in a most adverse way.

In the summer of 1965 I was summoned to see Sir George Edwards. Sir George showed me a letter from the Air Worthiness Authority complaining that they did not consider that the British Test Pilot input into the Anglo-French Concorde project was adequate. At the end of our discussions, I recommended to Sir George that it would be necessary to assign someone with recent Certification experience into the programme. Sir George stressed that whoever was chosen would have to be prepared to take the aircraft to M=2.0 (i.e. twice the speed of sound). As I left, I made it clear to Sir George that I was one very willing volunteer.

On 5 August 1965, some six weeks after our initial meeting, I was sent for again by Sir George. He told me that he had known what he wanted to do, but had to find a way of doing it. The outcome was my appointment as General Manager of Flight Operations of both the Weybridge and Filton Divisions of the British Aircraft Corporation with immediate effect. Sir George also asked me to go to Filton the same afternoon to see Archibald Russell (later Sir Archibald Russell) who was head of the Concorde Technical Team on the British side and Bob Brown, General Manager of Filton. I was thrilled to be faced with such a huge challenge and was in no doubt that I would have to do a great deal of catching up as the prototypes were in an advanced state of construction by this time – 001 at Toulouse and 002 at Filton.

I was also concerned whether or not there might be some resentment of an interloper from Weybridge. Rivalry dies hard and the formation of the British Aircraft Corporation from Vickers-Armstrongs, English Electric, Bristol Aeroplane Company and Hunting did not eradicate some sensitive feelings overnight. As it happened the welcome that I received from Russ (Sir Archibald Russell) and Bob Brown was extremely warm, so I knew that I could look forward to a long and happy association. Russ expressed some very strong views about our French partner who, he said, was determined to see the end of the British civil aircraft industry.

Russ had been a member of STAC chaired by Morien Morgan, Deputy Director at the Royal Aircraft Establishment, when it was formed in 1956. Consequently, he knew all about the trials and tribulations that had to be faced from the start of the programme. I met the Chief Engineer, Bill Strang and Assistant Chief Engineer, Mick Wilde within days of my first visit. I was given an enormous amount of help by these two brilliant engineers, who both became great personal friends.

I realized that I had a great deal to learn, not only about the aircraft itself but also regarding the organization behind the project. It was immediately clear that Concorde's system of management was extremely complicated. One has to remember that the Concorde was Europe's first, biggest and most complex collaborative project. The basis of the organization was six partners (two governments, two airframe companies and two engine companies) on a supposedly 50:50 split between Britain and France on costs, work and management. Not exactly the ideal way to run an advanced technology project and needless to say management by committee resulted. The peculiarities of the organization prompted Sir George to say in later years, 'These are not usually taken to be the ingredients for success'. He was referring to two languages, two traditions, two systems of measurement, two production centres, separated by physical and national barriers 600 miles apart. He went on to add that, 'These international programmes tried men's very souls'.

In simple, straightforward language, no single person was in charge. Recalling some of the paraphernalia attached to the project, it is a miracle that the programme actually happened. However, many of the shortcomings were overcome by the dedication of those involved.

In the early days of the Concorde programme the Government's direction was vested in the Concorde Directing Committee through to the Administrative and Technical Sub-Committee. This changed in 1968 with the formation of the Concorde Management Board. Over the period 1966–70 we were lucky to have Sir James Hamilton as Director-

General of the Concorde Division at the Ministry of Aviation, a man of great ability.

On the manufacturer's side, there was an Airframe Committee and Engine Committee, with the senior posts planned to alternate every two years. The initial members were: General André Puget, Chairman/Managing Director; Sir George Edwards, Vice Chairman/Deputy Managing Director; Pierre Satre, Technical Director; Sir Archibald Russell, Deputy Technical Director; J.F. Harper, Director of Production/Finance and Contracts; Louis Giusta, Deputy Director of Production/Finance and Contracts; A.H.C. Greenwood and W.J. Jakimiuk, Joint Sales Directors; G.E. Knight and B.C. Vallieres, Directors. On the engine side: Sir Arnold Hall, Chairman and Managing Director; H.A. Desbrueres, Vice Chairman/Deputy Managing Director; Dr E.J. Warlow-Davies, Technical Director; M. Garnier, Deputy Technical Director; R. Able, Director of Production; W.F. Saxton, Deputy Director of Production; J. Bloch and H.W. Rees, Directors.

Below all these figures came Dr Bill Strang and Monsieur Lucien Servanty. In addition there was a special team of six senior design engineers consisting of Mick Wilde, Doug Thorne and Doug Vickery from BAC and Gilbert Cormery, Etienne Farge and Jean Resch from Sud. All the senior people in Sud spoke fluent English. The only one who could not got himself into a slight tangle when he declared that he much admired the way that the British 'Had squeezed their teets' in the Second World War, when he was apparently intending to say 'Clenched their fists'! In typically British style, the same could not be said of BAC personnel and a way had to be found to bridge the language gap. Language courses were arranged by the personnel department but the man who did most to bridge the gap was Paul Cameron, whose personality and enthusiasm gained everyone's confidence immediately he appeared at the numerous joint meetings.

A further committee of airline vice-presidents was formed under the Chairmanship of Bill Mentzer of United Airlines in an attempt to reach common requirements for cockpit layout, choice of systems equipment and for everything else that affected airline operation, servicing and maintenance. This type of practice was used on the Boeing 747. Approval of the Committee of Directors was given for Bill Mentzer's Concorde Committee.

I well remember some of the cockpit meetings with all the Concorde option holders represented because I found them a nightmare. There had not been any airline input into the cockpit when the project began, so it was not surprising that some of the proposals were not liked by the airlines. Furthermore André Turcat was convinced that the Concorde required a moving map display, made by Sagem-Ferranti, who had produced a similar system for TSR2. It made sense for the latter but not for a civil airliner which was likely to spend most of its time over the sea because of the sonic boom. The map display occupied pride of place in the middle of the centre instrument panel on the prototypes but was thrown out by the airlines.

The operational approval for the first flight of 002 came from the Director of Flying, Ministry of Technology. He indeed issued MOD approval of the test pilots. Air Publication AVP 67, which defines all the rules and regulations for the flight testing of military aircraft, was modified to a form suitable for Concorde. This did not last long as a publication on behalf of the British and French authorities (Ministry of Aviation, Air Registration Board, Service Technique Aéronautique and Secrétariat Général A L'Aviation Civile) was produced in September 1970 with the intention of defining the procedures for both eventual certification and for the authorization of flight during the development phase.

This produced, firstly, the Concorde Airworthiness Committee (CAC) consisting of representatives of the four main constructors with their opposite numbers in the official services set up for coordinating the airworthiness and certification work. Secondly, the Joint Certification Committee (JCC) consisting of the same representatives that sat on the CAC with one additional member from the ARB. Thirdly, Flight Test were not left untouched. A Flight Test Committee (CDV) for the coordination of all the activities of the various official bodies with regard to the flight testing of Concorde was set up and initially consisted of M.C. Lenseigne, Vice-Chairman, 1971; M.J. Renaudie, Head of French Delegation; J.D. Hayhurst, Ministry of Aviation and Supply, Chairman, 1971 and H.C. Black, ARB. Chairmanship changed annually.

Under the CDV came the Flight Test Group (FTG): M.J. Renaudie, Head of French Delegation; M.A. Cavin, CEV; M.P. Dudal, Test Pilot. M.F. Gillon, Toulouse; H. J. Allwright, Head of British Delegation; J.C. Chaplin, ARB; S.G. Corps, Test Pilot, ARB; H. J. Keyes, Ministry of Aviation and Supply.

The CDV and the national head of the FTG held the right to delay the execution of any test until they were satisfied with the safety conditions proposed for that test. Procedures for authorizing first flights and for engine clearance before Certificate of Airworthiness were also detailed at some length. Finally the 'Grand Livre' procedure was detailed. The official definition reads: 'TSS Standard No. 0-2 requires the constructors to prepare a certification compliance register as evidence of compliance with airworthiness requirements. This document is called the "Grande Livre".' In fact the 'Grande Livre' consisted of a collection of sheets, each sheet recording compliance of a single system or a single zone of the aircraft with one requirement of the TSS standards.

Gradually, the top level became slightly more streamlined in 1968 after the Concorde Management Board was born and further changes occurred in 1971. There were constant changes of personnel on the official side at all levels. In Kenneth Owen's book *Concorde – New Shape in the Sky*, one chapter is entitled 'How not to get organized'. Being wise after the event is the easiest thing in the world, but the fact remains that Concorde got there.

Another important area that required particular attention was the matter of routes and the coverage of test flights by radar. On the British side the Director of Flying, Ministry of Technology organized the whole arrangement. The joint document was issued by the Director of Flying, Ministry of Technology and the Director of the French Flight Test Centre.

One feature which was of outstanding help was the use of a single VHF of 142.15 Megacycles throughout each flight from time of taxi to final engine shut down covering all the different radars both Military and Civil which were used in order to provide continuous radar cover.

Supersonic flying had to take place over the sea due to the sonic boom. Normally the air ahead of an aircraft moves out of the way up to the speed of sound, but as the speed of sound is exceeded, the aircraft arrives without warning and so the air particles build up in front of it producing a shock wave in the form of a cone rather similar to the bow wave of a ship. The intensity of the over-pressure diminishes towards the edge of the cone, which is about 25 miles wide. Wherever any section of the cone reaches the ground, a boom is heard. This is normally startling rather than damaging. The boom carpet is present all the time that the aircraft remains supersonic. However, during the acceleration through M=1.0, a focus boom confined to a very small area of about 100 yards or so is created by the arrival of several booms at the same point. The focus boom is much stronger. Flight trials revealed an unknown feature now known as the secondary boom, which is of much lower intensity than the normal boom. The secondary boom is the downward refraction under particular weather conditions of the up-ward travelling part of the cone and it occurs much further ahead of the aircraft, which is normally decelerated to be subsonic at least 35 miles before the coast.

One special route involved flying overland down the west coast of Scotland, Wales and Cornwall. This was the subject of much activity within the British Government. I had to go to the House of Commons to see the Minister, Anthony Wedgwood Benn with Jim Hamilton present in order to answer questions fired off by Members of Parliament. The Minister introduced me to his side-kick and then departed. I unfortunately did not catch his name. When the meeting was over the junior minister asked me if I would like a whisky and soda, which I gladly accepted. After trudging through some long corridors, we went through a door into a small snuggery. I was taken aback when the bar lady said, 'And what will you have my Lord?', then realizing I was actually in the House of Lords.

As soon as I saw the prototype, I had a major problem with the nose and visor arrangement. The visor was a metal fairing with a small window in order to streamline the nose section and to protect the cockpit windows from heat effects during cruise, where structural temperatures reach 127°C at the nose. The window itself did not give direct vision but was a mirror image. In fact when the visor was up, the pilot really could not see anything ahead of him. I was concerned at the prospect of having to land with the visor up in the event of failure to lower by normal or emergency systems. I was not helped by the fact that – by some strange act of bullshit – the Concorde prototype met the forward vision requirements agreed by the Airworthiness Authorities for supersonic flight. I was virtually led aside and told that this was a supersonic aircraft. In the end the matter was taken out of my hands by the US FAA who told me over a drink that they would never certify the aircraft with the existing visor as, apart from anything else, they would be crucified by ALPA.

When I reported this, a very different approach was taken and it suddenly became possible to have transparent panels which could stand the heat, made by Triplex. A long argument then started as to whether it should be Triplex UK or Triplex France. It so happened that the top of the Flight Operations Building at Filton was the same height as that at which the pilot's eyes are off the ground when the main wheels touch the runway at normal landing speeds – this is approximately 38 ft. The various panels used to be rigged up so that the test pilots could look at the runway lights at night in order to check that the optics of these very thick panels were acceptable. The British Triplex turned out to be better than the French Triplex and was chosen. The design of a new transparent visor, which is what the Concorde has today, meant a re-design of the whole movable nose, which is provided to give adequate vision for take-off and landing. This represented a major difference between the prototypes and the pre-production/production aircraft.

The second major problem was the Filton runway itself which had been specially made for the Bristol Brabazon, and was very much the pride of Filton. I found this difficult to understand because it was so much up and down that one could not see one end from the other; the slope was quite unacceptable for any measured take-off or landing tests and the first 1,000 ft landing towards the west was sterilized in order to keep aircraft high enough landing over the main A38 road. I did not consider that it was long enough for development flying on Concorde and the only good thing about it was the fact that it was twice as wide as any normal runway, which was unnecessary anyway. My criticism went down like a pricked balloon and I was confronted with masses of facts and figures showing what the prototype's take-off and landing weights could be with normal factors applied. This did not impress me at all because the fundamental common-sense requirement – that one needs extra margins for development flying of advanced prototypes – had been disregarded completely.

In the end, Sir Archibald Russell, Bill Strang, Mick Wilde, George Gedge and myself finished up in Sir George Edwards's office. Sir George listened to both sides of the argument and finally said, 'He is quite right'. None of us was ready to move the whole operation to Toulouse, especially as the flight testing pre-production/production had to be taken into account. George Gedge was in fact a supporter of my case and we had to decide where the British side of the flight test programme would be undertaken. Our minds turned towards the Government establishment at Boscombe Down, but after about two meetings to discuss the facilities required, it was quite clear that Boscombe Down did not want Concorde. It was a difficult journey by road anyway, so there was great relief when someone suggested Fairford. RAF Fairford had started life in the Second World War mainly for glider towing with Stirlings, was considerably extended for use by the SAC, and was then used for many years by the US Strategic Air Command with B-36s, B-47s and B-52s. The runway was 10,000 ft long with 1,000 ft each end of bearing surface and it was mainly flat. The question was could we get permission to use it as, by this time, there was one RAF C-130 Hercules Squadron there and they were due to leave. Our Vice-Chairman, Sir Geoffrey Tuttle, knew the Commander-in-Chief of Air Support Command, Air Chief Marshall Sir Thomas Prickett, very well and a meeting was arranged with him at Upavon attended by Sir Geoffrey and myself. Sir Geoffrey explained the predicament in some detail and after minor surprise expressed by Sir Thomas, agreement was given that the Concorde development could take place at Fairford, which was to become a relief station, but still under the auspices of the C-in-C Air Support Command.

Setting up the flight test operation at Fairford was then another major task to be completed before first flight of 002. A hangar, followed later by a second, with a Concorde weighbridge installed was made available together with office, administrative and technical accommodation while a special engine-running bay with silencers had to be constructed. Telephone links direct to Toulouse and a whole network of telecommuni-cations were put in place. A flight operations room with full radio and telephone links was established at one end of the technical office block. Domestic housing of flight test and design support and maintenance personnel had to be set in motion with suitable incentives to move. Much of the overall task was completed by Eric Hyde and John Dickens from Filton and we employed ex-Wing Commander Tony Pearson, whom I had known at RAF Dishforth twenty years before, as BAe's man on the spot to do all the liaison work with the RAF Station Commander and his staff at Fairford. It was an obvious worry that the first aircraft to see the new organization would be the Concorde 002, but that simply had to be faced.

The French did not stand still on establishing facilities either. Although the Toulouse airport had an existing 9,000 ft runway, a new parallel runway of 12,000 ft was built for the benefit of Concorde, although obviously available for use by other aircraft as well. The runways run 150°/330° (15/33). A crash barrier specifically designed for Concorde was installed at the end of runway 33 and an ILS (Instrument Landing System) was located at each end and a new production hangar was constructed next to the existing flight hangar. Toulouse has never looked back and has gone on expanding and expanding for Airbus until it is now as good a facility as one can see anywhere in the world. This is an illustration of a country with a Government determined to have an aircraft industry and one without any degree of consistency.

Partly due to André Turcat's position at Sud and partly due to the Fairford set-up, my title was changed to Director of Flight Test and Chief Test Pilot and I had responsibility for all personnel and operations at Fairford. At the height of the programme this was, roughly, 500 people. Setting up the Flight Test Organization was a task in itself because the degree of flight testing at Filton in the previous years had been rather limited. Consequently, the Filton Flight Test Department, although containing some very good individuals was not big enough and lacked recent civil aircraft experience. I brought in most of the Flight Test Department from Wisley under Mike Crisp with a number of stalwarts in Roy Holland, Mike Bailey and several others. Test pilots, John Cochrane, Peter Baker, Eddie McNamara and Johnnie Walker came into the programme immediately. The department expanded in all areas as the Concorde Flight Test Programme developed during the ensuing years. This also applied to the maintenance staff under Charlie Andrews from Filton and later Wally Chapman from Weybridge. Just like the Concorde itself, it was a moving feast which put a few noses out of joint in the process. Bob McKinlay came in as Assistant Director Flight Test in charge of Technical Staff and with a design office approval.

It had been intended to have two flight development simulators, one at Toulouse and one at Filton for the development phase. As an economy before I joined the programme, the Filton simulator was axed on the grounds that handling qualities was a French responsibility. The simulator was equipped with a three axis-motion system which had a serious limitation in that the effect of yaw had to be simulated by rolling the aircraft. The simulator could be connected to the flight control rig which was adjacent to it in the same building. As a result, Filton had a fixed base device, which was used for minor simulation work. John Cochrane, Peter Baker and myself spent a great deal of time on the simulator in Toulouse investigating Flying Qualities although it was a Sud responsibility.

BAC, on the other hand, were responsible for a major task on the simulator defined as Crew Workload and Studies, where Johnnie Walker and Eddie McNamara played the major role. It always struck me that this was a real cart-before-the-horse exercise because it was somewhat late to decide on crew work load when the aircraft and cockpit layout were already defined. However, it was an official requirement and had to be done.

We thought that we had a handling problem following engine failure on take-off. Because roll was used to simulate yaw, the wrong rudder was frequently applied to correct the effect of the engine failure and the aircraft finished up inverted. Fortunately this error was discovered quite early on and the cure was to turn the motion off for this particular test. The simulator was basically pessimistic and once the prototypes flew, it was possible to see how much easier the aircraft was to fly compared to the simulator. The simulator ran away on some occasions, whereupon a long delay occurred while ladders were fetched to enable the occupants to get out.

The next area to be addressed was the matter of flying practice. So much time was spent at meetings and learning the aircraft, that actual flying in anything other than BAC's communication fleet was at a premium. Furthermore, there was a general feeling between Director of Flying, Ministry of Technology and myself that a sharpening-up exercise should be organized for the lead test pilots at least. Through the good offices of the Director of Flying, flying in the Lightning at the Empire Test Pilots' School, in a Vulcan at B Squadron Boscombe Down, a Hunter at Farnborough and the BAC221 and Handley Page 115 at the Royal Aircraft Establishment Aero Flight at Bedford was arranged. The BAC221 was one of two Fairey Deltas modified with an ogee wing while the HP115 was a small slender delta with the simplest of systems built for low-speed investigations. The FD2 airframe converted to the BAC221 was the aircraft that held the world speed record when flown by Peter Twiss on 10 March 1956. We were also offered flying in a Mirage III at Bretigny and Cazaux in France. A Mirage IV was attached to the Sud flight test department which clearly illustrated the priority that existed towards Concorde. The Mirage IV was a M=2.0 Bomber and belonged to the elite Force de Frappe.

I remember my first trip in the Mirage IV from Toulouse

very well. I was shown the cockpit by Gilbert Defer, who was still in the French Air Force prior to his joining Sud. He was smoking a cigarette and the ash kept falling into the cockpit. The other crew member who occupied the rear cockpit was M. Cavin of CEV who spoke very little English. There were no restrictions on flying supersonically up to Mach=2.0 over France, which was a bit surprising. It was a delightful aircraft to fly and we all thoroughly enjoyed the experience, although I was quite nervous about flying this aircraft with such a basic understanding, especially as my ability to communicate with M. Cavin was almost zero. Unfortunately, the aircraft crashed later on when being flown by one of the French official pilots, but both crew members ejected safely.

In addition I had the good fortune to go to Edwards Air Force Base, California for flight experience in the Convair B-58 supersonic bomber thanks to our Controller of Aircraft, Air Chief Marshal Sir Christopher Hartley and the Commanding General, Al Boyd of Wright-Patterson Air Force Base. The Officer Commanding Bomber Test Squadron at Edwards was Colonel Joe Cotton, a graduate of the Empire Test Pilots' School whom I had met when he was there. He recommended that there was no point in me turning up unless I was allocated ten hours of flying on the B-58, which duly happened and was of unique value. The B-58 was rather a handful and had suffered a number of accidents following an outboard engine failure at M=2.0. The aircraft was equipped with auto-stabilization (stability augmentation system – SAS to use the USAF term) on all three axes. I was carefully briefed to switch the SAS off before becoming supersonic, pulse the aircraft and then switch the SAS on in order to prove that it was working. I was reminded that 'If an outer engine fails at M=2.0 without the SAS working you're been', a polite way of saying that the fin would come off.

I made four flights altogether with Joe Cotton and his deputy Major Ted Sturmthal. The B-58 carried a crew of three seated in separate compartments behind each other, and I did some landings from the instructor pilot's position, i.e. the number two compartment, which was quite a revelation. It was very difficult to see the runway from this position at normal approach speeds around 160–70 knots, so the technique adopted was to increase the final approach speed to 225 k, thus flattening the aircraft's attitude in order to see ahead. The ten hours took nearly one month, as it was necessary to change one of the engines after each flight due to unforeseen defects. I filled in the time between flights flying in the B-52 that was used for carrying research aircraft like the X-15. I will always treasure the warm welcome I received at Edwards.

Back in the UK, thought was given to providing a chase aircraft for Concorde. The best that could be spared was a rather ancient Canberra B2, which was also available for continuation training, and was used on all the early flights of 002.

TWELVE

Concorde: Some Technical Aspects

Work sharing on collaborative projects is never easily agreed, although on Concorde the principle of a 50:50 split in terms of cost played a major role in resolving this issue. On structures, the French were responsible for the wing, the centre fuselage, the elevons and the landing gear. The British were responsible for the forward fuselage including droop nose/visor, engine nacelles, the rear section of the fuselage, the fin and rudder. On systems, the power plant (i.e. nacelles, engine controls, fire warning and protection), fuel system, electrical generation and distribution, oxygen, thermal and sound insulation, circulation and distribution of cabin air were British responsibility. The French had hydraulics, flying controls, automatic pilot, automatic stabilization, radio and navigation, generation control of air conditioning. In aerodynamics, the British had responsibility for aircraft performance in all aspects and de-icing while the French were in charge of handling qualities. The flight test programme was supposed to be drawn up to reflect the design responsibilities by allocating tasks to each aircraft accordingly. This worked to some extent in that BAC did most of power plant and performance testing while Sud (SNIAS 1970) concentrated on handling. Nevertheless it was necessary to be reasonably flexible in the allocation of tasks.

The Concorde design teams had no past practical knowledge of a supersonic airliner, so they had to draw on recorded data from a number of military or research aircraft.

In France there was the Trident, the Griffon, the Durandal, the Mirage III and the Mirage IV, while in the UK there was the Lightning, the Vulcan, the Bristol 188, the Fairey Delta FD2/BAC221, the Handley Page 115 and TSR2. Future generations of supersonic civil airline design teams will never have to face these problems because there will always be Concorde and the Russian Tupolev Tu-144 knowledge to draw on. I believe that the two most difficult design problems were to decide the shape in the first place and secondly getting the engine/intake to work efficiently.

Rolls-Royce (originally Bristol Siddeley) were responsible for the Olympus 593 engine. Their brief was the design, development and manufacture of the basic engine (i.e., the compressor, combustion chambers, turbine, primary nozzle, fuel control system and accessories except reheat fuel control system). While SNECMA would design, develop and manufacture the rear end of the engine, i.e. reheat fuel control system, variable convergent/divergent nozzle and jet pipe with a noise suppresser and thrust reverser and the reheat control system. An Olympus 593 was installed in the bomb bay of a Vulcan bomber in order to prove the power plant at subsonic speeds prior to the first flights of the prototypes. I found this to be a very valuable way of learning something about the engine handling characteristics and re-lighting capabilities.

One of the worst aspects of Concorde development history relates to the selection of equipment for the aircraft systems. Whereas the airframe and engine companies had their work share specified from the start, the equipment choice was wide open. There was a great deal of political manoeuvring and horse-trading between the two sides and the final choices were not always decided on pure technical merit. A lot of the jockeying for position had already taken place before I joined the programme but I observed plenty of it as decisions for the production models were made.

The French undoubtedly saw Concorde as a splendid

opportunity to improve the status and standing of their equipment industry. The British equipment and materials manufacturers became greatly concerned and made their views known to the Government through the SBAC and the EEA. They were particularly concerned at the possibility of American equipment being manufactured under licence in France. Some of the areas which caused most strife were the powered flying controls, main generators and hydraulic pumps.

Concorde had a number of special features that made it different from anything that had gone before in the Western world. Firstly, at take-off and landing speeds, the Concorde assumes very large pitch angles, which is typical of delta wing configurations. In order to provide adequate forward vision for the pilots a movable nose and visor were incorporated. For taxi, take-off and initial climb, the nose is set to 5° with visor down; for approach and landing the nose is lowered to 12.5° (originally 17.5° on the prototypes) with visor down. For climb and cruise the nose is raised to 0° with visor up. As already mentioned, this was one area of major re-design and appeared on the first pre-production aircraft. Secondly, in order to provide air into the engine that is digestible and to minimize airflow distortion at the engine face, a variable geometry intake system consisting of two automatically controlled ramps is provided.

The purpose of the intake is to slow the incoming air down to a subsonic speed of M=0.5 over a length of 15 ft by means of a shock wave pattern which is set up by the movable ramps. Each intake had also an auxiliary inlet/spill door on the bottom side. Any sudden breakdown of the airflow into the engine will cause a 'surge'. This is like a forward-firing belch from the engine akin to an explosion. The intake and the nozzles contribute over 90 per cent of the total thrust in supersonic cruise at M=2.0.

Military aircraft normally fly for relatively short periods at supersonic speeds and can afford to have plenty of margin between normal engine operation and surge. In the case of Concorde this was not possible because it was vital to squeeze the maximum range possible. Consequently, the margin between normal operation and surge is set to the minimum. It was equally essential to produce a system that gave surge-free operation throughout the normal flight envelope.

The development of the power plant required a number of major changes both physical and in the control laws. In some ways this did not altogether surprise me because I always thought that pushing air into a rectangular box, which then had to find its way into a round hole was rather a peculiar way of doing things. The situation was further complicated by the fact that the four intakes are different, although one might expect all four nacelles that appear to be pointing in the same direction to be the same. They do not point in the same direction exactly because the two pairs of intakes are toed in at a slight angle due to the effect of wing sweep. There are also slight differences in the aerodynamics at the inboard and outboard positions. Thus, although the engines all rotate in the same direction, airflows at the entry into each engine are slightly different. In-flight experience necessitated many alterations, which are addressed in chapter thirteen.

The third unusual feature is that the centre of pressure on a wing moves aft in supersonic flight, thus upsetting the balance of the aircraft and causing a nose-down trim change. If the centre of gravity is left in a fixed position, the pilot would counter the trim change by pulling back on the stick, thus applying, in the case of Concorde, up elevon. This would cause an excessive drag penalty and could not therefore be tolerated. Although the variable camber and twist of the wing gave some alleviation, the real cure was to move the centre of gravity by pumping fuel backwards and forwards to suit the particular flight condition. The fuel system is designed to permit this and the range of c.g. position for a particular Mach number is displayed to the crew by means of bugs on the Machmeter. The pilots and flight engineer are also provided with a c.g. indicator which shows the c.g. limits for the speed being flown. Control of c.g. is carried out by the flight engineer.

Fourthly, the Concorde flight control system uses electrical signalling between the pilots' control column and the power control unit at each surface with a mechanical linkage as an emergency back-up. Fifthly, in supersonic flight, an aircraft's structure is subject to thermal heating due to skin friction. In Concorde's case the nose reaches 127°C in cruise. Because of this phenomenon, the fuel system is also used as a heat sink for the air-conditioning system.

The decision to use aluminium alloy for the structure of Concorde was directly related to the choice of cruising speed. Originally it was planned to cruise at M=2.2 or in temperature terms 425° Kelvin. Early calculations showed that a reduction to M=2.05 and 400°K would vastly improve fatigue life while only giving a minute flight time penalty. There were already enough unknowns to be covered in developing a supersonic airliner without embarking on new materials of which there was only limited experience. So the Concorde became and is an aluminium alloy construction, although some titanium is used in the power plant. On the manufacturing side, BAC built the front and the rear parts of the aircraft and the engine nacelles while Aerospatiale built the middle and the wing. These components were shipped by road and sea, sometimes by air between the two assembly lines at Filton and Toulouse.

As might be expected, there were numerous structural test specimens and rigs including a full-scale fuel rig capable of attaining any of the attitudes that might be encountered by Concorde. The hydraulic/flying control rig at Toulouse has already been detailed in the last chapter. Two complete Concorde airframes were manufactured to cover fatigue life testing at RAE Farnborough and structural strength testing at Toulouse. The former was placed in a specially constructed facility at Farnborough in order to simulate the actual loads and stresses brought about by the heating and cooling that occurs in each flight cycle of subsonic to supersonic and back to subsonic flight.

On the sales side, prospects looked quite good. Pan Am placed the first provisional order for three, which was then increased to eight. There were seventy-four options from sixteen major airlines by the end of 1967. Gradually, as the years wore on this finally dwindled to sixteen production, two prototypes and two pre-production aircraft. Towards the end, a troika of BA, Air France and Pan Am existed but Pan Am finally pulled out. Various reasons have been quoted for the withdrawal: cost played a part, but the lack of permission to land at New York and the sudden rise in fuel prices sealed the lid very firmly. Eventually BA and Air France took seven each and the first two production aircraft never entered airline service.

One of the most difficult areas to be considered on civil prototypes is the matter of crew escape. From a psychological point of view at least, there is a case to be answered even when one's instinct knows how difficult it is to reach an escape hatch some distance from the various crew positions. The presence of escape hatches on the Concorde prototypes introduced an additional problem. It was calculated that blowing off a hatch at 60,000 ft at M=2.0 would initially take the Concorde cabin to about 72,000 ft due to suck. Such an altitude is outside the capabilities of oxygen pressure breathing equipment. Consequently, we had to face up to the use of partial pressure suits and helmets and all those involved in flying the prototypes were trained at the RAF Institute of Aviation Medicine in the use of this equipment. Immersion suits were worn as well because chances of survival without one in the event of coming down in the North Sea were nil. I found all this 'clobber' most uncomfortable in a cockpit which had not been designed for its use and I decided quite early on that I could not see some of the essential switches and instruments when wearing the pressure helmet, so I ditched the helmet on the grounds that I was nothing short of a menace with it on and the pressure mask would have to suffice.

We used to practise crew escape from the wooden cockpit mock-up. It was especially difficult getting out of the co-pilot's seat. The captain's was easier because his seat was motorized fore and aft and up and down. On one occasion I was in such a hurry that I forgot to undo some of the connections. In my efforts to get free I nearly destroyed the mock-up. For all the early flying, full safety equipment was worn using a back type parachute and life jacket (Mae West) and sitting on a dinghy pack.

THIRTEEN

Concorde: the Flight Test Programme

In attempting to describe the flight test programme it is important to understand that the definition of the aircraft for airline use changed regularly for several years in order to produce a viable proposition. Sir George Edwards once said, 'We had to run very fast to standstill'. A great deal of the changes resulted from wind tunnel and other ground tests and only a few mainly associated with the power-plant engine intakes came from the prototype flying. Furthermore, the airworthiness regulations, TSS, had to be developed as well.

From a flight test point of view a number of critical tests had to be repeated for each variant. I recall three or four in-flight flutter programmes, prototypes, pre-production No. 1, pre-production No. 2 and production. Even production aircraft Nos 1 and 2 were not 100 per cent representative of the in-service aircraft.

Segration of aircraft systems against the possibility of serious failures like an engine disc burst that was not contained were unacceptable on the prototypes and required more than one modification. The first pre-production aircraft was equipped with a pseudo-auxiliary power unit (MEPU) which used hydrazine fuel. Hydrazine fuel was thoroughly nasty stuff to handle on the ground, so the maintenance crew had to wear protective clothing during refuelling operations. An unlikely choice for a civil airliner one might say. On the prototypes two GTS were fitted to engines 2 and 4 for engine starting and for driving the auxiliary gear boxes.

This was another case of poor thinking because if a GTS failed to start there was no alternative way of starting that particular engine. What the airlines wanted and got in the end was a conventional air start system from a ground truck with crossfeed between each pair of engines. A flight test programme using the two prototypes, two pre-production, production 1 and 2 and route proving on aircraft 3 and 4 was drawn up.

It is not historically unusual for prototypes to fly later than originally anticipated; it has happened before and it still happens today. The original concept for Concorde was hopelessly optimistic – first flight of the first prototype was scheduled for the second half of 1966, first flight of first production aircraft the end of 1968 and Certificate of Airworthiness the end of 1969.

The first appearance of a Concorde took place on 11 December 1967 at Toulouse when 001 was rolled out on an incredibly cold but sunny day. This was a big occasion attended by Monsieur Jean Chamant, Minister of Transport and Anthony Wedgwood Benn, Minister of Technology as well as about 700 guests flown in from Britain and Paris. The ceremony started with nearly an hour-long speech from Monsieur Chamant followed by a rather shorter one from our Minister. Sir George Edwards and Monsieur Papon, the ex-Chief of Paris Police, who had replaced General Puget as Head of Aerospatiale, also spoke. This was the occasion when Wedgwood Benn announced that the UK would in future spell Concord with an 'e', which went down very well. A French Air Force Band played 'God Save the Queen' after the RAF Band specially flown over for the occasion had played the 'Marseillaise'.

The completely frozen onlookers, sitting in a stand constructed in front of the hangar, breathed a huge sigh of relief as the hangar doors opened and 001 was towed out into the sunshine. As 001 vacated the warm hangar, it was filled by the guests who immediately descended on the champagne and smoked salmon. I felt particularly sorry for the air stewardesses lined up in front of the stand, representing each of the sixteen airlines holding options to buy Concorde, dressed in their flimsy uniforms. According to Reginald Turnhill, the press were badly done by in the way of facilities. Security thanks to M. Papon was very tight and attendance was very restricted.

I had arrived in Toulouse the day before and was most impressed by the appearance of 001 both inside and out, as I had seen it a week or so before when one might have thought that a bomb had gone off inside. Wires were hanging everywhere and only a few instruments were in place. I was amazed by the transformation in so short a time until I found out, by trial and error, that none of the instruments or wires were connected up, as they all had been pushed into their respective holes in order to produce what was an immaculate flight deck and rear cabin. A flight date of 28 February was still maintained throughout the proceedings – the only trouble being that it was 1969 not 1968.

When 002 was rolled out at Filton nine months after 001, there was no ceremony, no guests, only the work force to watch it. It was regarded as a normal working day and the aircraft was taken straight to the running base for commencement of ground runs. I believe that Minister Wedgwood Benn was furious that he had not been asked and even the Director-General of Concorde DTI Sir James Hamilton had not been informed.

In August 1968, 001 commenced taxi trials. I accompanied André Turcat as co-pilot, although we had already decided between us that we would not fly high-risk flights with mixed crews. But for the taxi tests, it was something of a political move, although for me personally it was a more than useful experience for the future.

All the ground-running tests on 002 were carried out by a full flight crew. This proved to be essential so that all the aircraft systems were set up properly, otherwise damage could occur. A very elaborate engine-running base with an underground bunker was constructed at the west end of Filton with large de-tuners and baffle fences. The operations room in the bunker was equipped with closed circuit TV and radio communications, so that every form of contact with the aircraft could be maintained by the large group of technical staff who manned the bunker. It was necessary to carry out some engine running out of de-tuners to get truly representative resets. The baffle fences did not care for this and disappeared into the adjoining fields on the first occasion. I managed to get permission to taxi 002 from the running base back to the hangar where it lived towards the end of the ground-running trials.

Towards the end of 1968 it seemed that 002 could be ready for flight before 001 and a major exercise took place to see what could be left off 002. I had to veto some of the ideas – for instance, I was asked to accept only one radio set. The intent of flying 002 first, resulted in the installation of a crash barrier at the west end of the runway as it was felt that the French would never agree to 002 flying without one. In truth, the whole concept was flawed as the French were never likely to agree to 002 flying first and an agreement that 001 would get priority for any spares put an end to this possibility.

The huge differences between the prototypes and the production aircraft had by this time been assimilated by everyone. The prototypes were really by now a proof of concept and the number of demonstration flights and tours which took place throughout the early part of the flight test programme reflect this philosophy and the need to keep Concorde in the public eye worldwide. The data acquisition system of flight-test instrumentation was capable of measuring approximately 4,000 parameters and practically filled the passenger cabin, where it was monitored by three, sometimes four, flight-test observers. Its weight was such that the prototype could not land at its design weight with any normal amount of fuel on board.

The first flight of a supersonic civil transport was in fact that of the Russian Tu-144 under great secrecy on the last day of 1968. The Concorde's turn came on 2 March 1969 when André Turcat lifted 001 off runway 33 left at Toulouse for the first time. The glare of international publicity played a major part in the time and date of the first flight. It was decided that the first take-off must be made towards the crash barrier on runway 33 left, but for two days the wind was blowing from the opposite direction and the weather was misty. On the third day the visibility had improved but the wind was still blowing in the wrong direction.

It became known that there was a hideous problem with the hotel accommodation in the Toulouse area and from that point of view, flight was essential. This put enormous pressure on André Turcat and General Henri Ziegler, the Head of Aerospatiale made a number of visits to André's office which I was also using. André gave his decision to fly at about the middle of the day. It took at least one and a half hours to carry out the lengthy checklist which was initially used on the prototypes, but, in due time, all four engines were started and 001 taxied out for take-off on runway 33 left accepting a downwind component, which although not ideal, was considered to be quite safe.

Accompanying André Turcat was Jacques Guiguard as Co-pilot, Michel Retief, Flight Engineer, Henri Perrier, Chief Flight Test Observer, Claude Durand, Flight Observer and Jean Belon, Assistant Chief of Flight Test for SNECMA. Jacques Guiguard retired soon after the first flight for health reasons and we were all very sorry to see him go, as he was a great character, full of charm and fun. Henri Perrier would later succeed André as Director of Flight Test. I always think of Henri Perrier along with Robert McKinlay, who joined our flight test team at Fairford as Assistant Director of Flight Test, as the two people to be credited with making the greatest contributions to the flight development of Concorde.

The first flight of 001 lasted for forty-two minutes covering general handling and pilot's assessment up to 10,000 ft and 250 k before coming in for a perfect landing by André Turcat in spite of the down wind component. The whole of aviation was thrilled by the beautiful sight of a Concorde in the sky. Eight more flights were made during the month of March covering increases in speeds of up to 350 k and 30,000 ft and including assessments of handling in abnormal configurations, engine handling and relighting and performance in climb and cruise.

André very generously asked me to fly 001 on flight five on 21 March 1969, which I regarded as being an outstanding example of our solidarity and I respected him enormously for this. It was a thrilling experience which I enjoyed very much and it put me in good shape for flying 002 for the first time on 9 April 1969.

The last three days before the flight were very frustrating. Each time I accelerated the aircraft up to about 100 knots, the failure flag came up on the captain's airspeed indicator. Design office and test services tackled the fault on each occasion but did not cure it. The weather was perfect for the flight to Fairford and I knew that it was going to change as the first signs of the approach of a warm front appeared. On the third day, I decided to do another fast taxi check but had at the back of my mind that if the snag had been cleared I was going to keep going instead of returning to disposal to tell everybody that it was all right. This was indeed exactly what happened.

A Concorde take-off from Filton was quite an exercise in itself because the main A38 road had to be closed by a combination of traffic lights and police as it was necessary to use the whole runway length. This meant blasting the A38 in no uncertain manner as on the prototype the procedure was to open up to full engine power against the brakes and then after a few seconds select the re-heat (after burner) which preferred the engine gas temperature to have risen to near maximum. This was the best way to ensure re-heat light up.

It took over an hour to go through the pre-start check list and then it was a relatively short distance to taxi to the take-off point on the runway. I had told Air Traffic Control of my intentions, so I received take-off clearance before moving. One of the re-heats took two selections to light up and then we were on our way. The whole crew sighed with great relief as 002 passed 100 k without any failure flag. The acceleration during take-off was very high, as the aircraft weight was low. A fast rotation was required under these conditions and 002 was airborne for the first time. The crew was strangely quiet in comparison to the massive cheers from the large crowd around and on the airfield, which we subsequently saw on television.

At about 1,000 ft I turned right on to a north-easterly heading towards Fairford. Filton radar handed us over to Brize Norton who monitored most of the flight. Tests were confined to general handling as the main purpose was just to get to Fairford. At the time we joined the ILS for landing on runway 27, a light aircraft apparently got in our way and a 'near miss' was filed by Brize Norton, but I did not see any of this and was totally unaware of the light aircraft's presence. Then as we turned on to our final approach both radio altimeters started playing up and failed.

This was definitely unfriendly. The radio altimeter measures height above the ground very accurately and its use was and is standard procedure for the last part of any approach. Height calls over the last 100 ft or so being made by the co-pilot or flight engineer because, due to the high nose up altitude of Concorde during approach and landing, the pilots' eyes are 38 ft off the ground as the main wheels touch the runway. I was left with no alternative but to eye-ball the landing and make the best of the situation. We came in on a 3° glide path and I kept the power on until the main wheels touched the runway. The landing was firm but not too bad, although I was pretty fed up with the performance of the radio altimeter.

The landing roll out and taxi into the dispersal were straightforward but I noticed two particular features. Firstly, when the nose was lowered to $17\frac{1}{2}°$ for the landing it was rather like looking over a precipice as there was no reference in front of one's eyes. This setting was later changed to $12\frac{1}{2}°$ when one could see a bit of nose. Secondly, the high nose up attitude on the approach (about 11° pitch attitude) gave the impression that the aircraft was going to land half-way down the runway. Some pilots found this rather disconcerting and adopted a tendency to get too low on the approach, which is a highly dangerous practice because, if unchecked, the undercarriage would remain behind before the runway threshold was reached.

We received a warm reception from Sir George Edwards, who had flown from Filton ahead of us, and other directors of BAC and Government officials. There was the usual sea of photographers but there was not too much for them to get excited about because the flight had gone very well, with the exception of the radio altimeters and the near-miss report.

After de-briefing the whole crew went back to Filton where there was a sizeable gathering of press and other interested parties. The late Sheila Scott (of world-record fame) had baked an enormous birthday cake for 002 covered in cream and icing sugar – not very good for the figure. We then went to the Directors' Mess where each crew member received a present from the Directors. Finally, we met up with many of the chaps who had actually assembled 002 in one of the local hostelries and so ended a memorable day with a thick head to look forward to.

002's first flight lasted for twenty-two minutes flight time and forty-three minutes block time. One major problem which was particularly obvious, apart from noise, was the cloud of black exhaust smoke from the engines, especially at full power. Some action had to be taken and as a result a change of combustion chamber design was specified by July 1969.

001's main task was to clear the flight envelope for higher speeds and altitude. An extensive flutter programme especially in the transonic region was required. Flutter is the term given to a sustained oscillation of the structure. The purpose of flutter testing is aimed to demonstrate that any induced oscillations damp out thus showing positive structural damping. This is done by applying rapid movements of the control surfaces by artificial means followed by immediate release. The length of time it takes the resulting oscillation to die out is recorded and so are any other structural vibrations resulting from the initial input. This is commonly referred to as ensuring that 'the tail does not wag the dog'. For Concorde the control surfaces were fitted with small rocket packs, known as 'bonkers', and an out of balance exciter motor was fitted to the flight control hydraulics. An electronic stick jerk box was fitted which controlled the input very accurately and consistently instead of relying on jerks applied by the pilot, which often vary.

On 16 March 1969 002 flew for a second time, but a minor fault on the landing gear curtailed the flight. There then followed a series of flights covering envelope expansion, functioning of the nose/visor, landing gear and all systems, engine relighting, various emergency cases and an assessment by the CDV pilot, Gordon Corps, as well as aircraft climb and cruise performance measurements. Operation in heavy rain and on a wet runway were also covered. All subsonic flying was carried out with the nose at 5° and visor down. This produced a high noise level in the cockpit and visible vibration of the nose/visor structure. When the visor was raised the noise level in the cockpit was so low that it was almost uncanny.

During this phase of flight tests, control of the aircraft in mechanical mode, that is with both the electrical signalling systems switched off, was explored in some detail. It was considered to be satisfactory for a major emergency condition but the actual landing had to be executed rather carefully. There was a distinct lag between pilot pitch input and aircraft response, which led to a tendency to over-control and set up an oscillation in pitch.

On 7 and 8 June 001 and 002 both appeared in a fly-past display at the Paris Air Show, Le Bourget. We had not been able to practise a combined display but the appearance of a Concorde from each end of the display line caused a considerable stir. I demonstrated the air brakes, which were fitted on the rear fuselage on the prototypes on one run and forgot to retract them until after the next run. This did not matter because their effect at low subsonic speed was negligible.

No sooner had the Paris Air Show finished than 002 was required to take part in the fly past over Buckingham Palace on the occasion of the Queen's Birthday celebration. Special clearance was given to fly over London at 1,000 ft but the weather, although good, was very hazy and I had great difficulty in lining up with the flag-pole at Buckingham Palace. Fortunately, London Radar were superb in their assistance and I have a photograph of Concorde right over the top of Buckingham Palace, signed by the Queen to prove it.

Flights then continued for the rest of June, July and early August within an envelope of M=0.9 and 31,000 ft, covering handling, performance, structural measurements and intake behaviour. All take-offs and landings were always measured for distance. At the beginning of August, 002 commenced a long grounding which lasted eight months. André Turcat flew 002 twice on the day prior to the grounding in order to compare it with 001 and, happily, he thought there were no differences. Sir George Edwards, in full flying kit with parachute, flew with us and was the first non-member of flight test to do so.

Simulator tests had showed that at M=1.7 following outer engine failure, the aircraft was uncontrollable due to excessive roll and sideslip. The trouble was traced to the elevon gearing which was set at a 1:1 ratio between roll and pitch. This was applying a side force to the fin and thus setting up the condition. It was decided by the aerodynamicists to change the ratio to 03:1 but this meant a complete re-design and manufacture of the mixing box in the flight control system. A great deal of updating of all the other systems was included and an Inertial Navigation system was commissioned and different radio altimeters were fitted. The flutter excitation system was also activated.

Six flights followed the grounding. On March 25, 002 achieved M=1.0 for the first time. Additional flutter testing took place up to M=1.35 using the system already described. Aerodynamic data, engine handling tests, intake control tests and structural measurements were also obtained at different ranges of centre of gravity.

On flight thirty on 10 April, the Minister of Aviation, Anthony Wedgwood Benn, came on the test flight which was planned to be a fairly gentle affair. One test to be made was switching off one hydraulic system to the flight controls. Immediately I did this, the aircraft rolled to the right and I could not stop it. In a second or so I had switched back to normal and regained full control. The Minister was totally unaware, but I realized that a serious design fault existed. Investigations on the ground soon confirmed that a key valve in the flight control system had a design fault, which clearly had to be rectified before any further flying on either 001 or 002 took place. This took four months to rectify by re-design and re-testing on the flight control rig at Toulouse

During August 1970 we were back in business testing out the air intake control system, more flutter work gradually increasing speed, aircraft performance, engine relighting and appearing at the SBAC Flying Display at Farnborough. On the last day of the show the weather at Fairford was atrocious and I knew that the chances of landing back at Fairford were pretty slim but the weather at Farnborough was good and that at Heathrow it was even better. I made two attempts to land at Fairford after our display, without success, so I elected to land at Heathrow somewhat overweight on runway 28 left (now 29 left). At this stage, the approach and landing technique was to be established at target threshold speed from at least 800 ft downwards. This combined with the overweight meant use of quite high engine power all the way down the approach causing hundreds of noise complaints. A very light-weight take-off the following day using noise abatement technique made some amends, as I recall being nearly 2,500 ft by the end of the runway.

Speed/altitude was further increased in carefully controlled steps with particular attention being paid to simulated double-engine failure around M=1.7 with the automatic rudder system switched on and off. The effect in all cases was very innocuous and clearly demonstrated that the simulator worries had been eliminated by the gearing change on the elevons. In fact 001 was never modified, as the actual flight results were nothing like as bad as the simulator had suggested. As a result the gearing was set at 0.7:1 on the production aircraft. The auto-rudder was also shown to be unnecessary.

By November, the stage was well set for reaching M=2.0. André and I decided that 002 should make the first excursion because it had been modified and that 001 could follow immediately afterwards. Our first attempt was thwarted by a false fire warning on No. 2 engine. A second attempt five days later was also thwarted by loss of oil on No. 4 engine. In the meantime André was raring to go and this he did after our first abortive attempt on 4 November 1970 reaching M=2.0 without any difficulties.

Third time lucky, and 002 reached M=2.0 on 12 November 1970. This flight routed subsonic to the Wash, followed by acceleration up the North Sea, round the north of Scotland and then southbound on the special route over western Scotland and Wales heading for Cornwall.

Achievement of M=2.0 at 50,300 ft was absolutely trouble free with the whole aircraft working perfectly. At 20 miles per minute the route down the west coast did not take long but it was sufficient for our purpose. Just before we reached Cornwall, I thought that enough time had been spent in this condition, so I asked John Cochrane to gently throttle back for deceleration. At that moment we all thought a third world war had started as Nos 3 and 4 engines went into what can be described as cyclic surge (a series of several continuing surges). There were about seven interactive surges (surge from one engine affecting its neighbour) between engines 3 and 4. Having had no previous exposure to engine surge on Concorde, all the crew were somewhat surprised to say the least, but the surging stopped as speed reduced and we turned to Fairford.

A longer than normal de-brief ensued and the lengthy analysis of the flight tapes commenced. The next few flights concentrated on investigating the surge problem at various engine and speed conditions, resulting in a number of successfully induced surges, although some test conditions did not produce the expected surge. Even surges which were purposely induced, and in spite of feeling rough running prior surge, were unpleasant.

The fact that the Concorde intake was designed as a self-starting system or perhaps better described as a 'non-unstarting system' was a major blessing. Self-starting means that the engine does not lock into a surge condition, necessitating engine shut down as the only way of stopping it. The Lockheed SR71 Blackbird suffered intake 'unstarts' which have been described as being pretty horrific and violent causing the pilot's head to be shaken so much that his helmet was hitting each side of the cockpit. I do not believe that anything other than a self-starting intake is acceptable on a civil airliner and this may well limit the speed of the future SSTs to M=2.2.

The analysis of these early flights showed up several

deficiencies which needed correction. Engine throttling, application of negative g (force of gravity) or sideslip produced surges and this was an unacceptable situation. Fortunately, the engine throttling problem was overcome on the prototypes by the use of deploying the speed brakes, which were fitted to 001 and 002, when decelerating from M=2.0 and the engine throttles were not retarded until M=1.7. Alternatively, surge could be avoided by slamming the throttles shut but this produced an uncomfortably rapid deceleration. The initial surge behaviour meant that Concorde was not meeting the original objective of having engine handling that was at least as good as subsonic aircraft. Furthermore the fact that one engine upset its neighbour meant that Concorde could only be regarded as a four-engined aircraft up to M=1.6 and thereafter as a two-engined aircraft up to M=2.0.

Other problems were finding a suitable position to locate the sensor of the rather limited analogue computer used to control the movement of the intake ramps. Later in the programme and much against the wishes of the French, this arrangement was discarded and an improved digital computing system with a different sensor arrangement was adopted. I am convinced that this course of action saved the day and produced a satisfactory intake control system on the production aircraft. I often used to thank God for the black boxes that pushed out electronic signals to the hydraulically operated moving parts as I knew that if mechanical gears and rods had been used we would still be trying to make it work.

One particular major incident occurred when 001 experienced a very heavy surge at 530 k M=2 over the Atlantic caused by the cancellation of re-heat on No. 3 engine which resulted in an overspeed and surge on No. 3 engine, which interacted into No. 4 engine which surged so violently that the front intake ramp was spat out which damaged the lower lip of the intake. Various bits of metal were ingested by both Nos 3 and 4 engines. Gilbert Defer had no trouble handling the aircraft, but the large number of individual warning gongs and lights made communication difficult. All of this resulted in structural strengthening of the ramp mechanism and a change to the control laws. A significant reduction in the number and type of individual warnings was also made.

Clearance of the high speed end of the envelope was by no means the only activity. The low speed side was investigated very fully by 001. Concorde does not stall in the conventional sense but develops a very high sink rate as angle of attack is increased. 001 was flown down to 119 knots and with full power applied still descended at 7,000 ft/min. with wings level and reached a maximum of 28° angle of attack when directional and lateral control became very marginal.

Wind tunnel and free model tests had indicated that above 23° directional stability and elevon effectiveness would be reduced due to changes in the wing vortex pattern. Above 30° the models used to go wild, pitching up to about 60°, then becoming inverted before recovering and thundering back to 30°.

It was seen that an anti-high angle of attack system was required which consists of angle of attack trim (automatic trimming of elevons as angle of attack increases), which is part of the auto stabilization system operating at 11° and above, and a stick shaker, which activates at 16°. This is signalled from either of the two Air Data Computers and operates on the captain's control column but is also felt on the co-pilot's control column through the mechanical linkage, and an anti-stall system augments the basic pitch auto-stabilization with a super-stabilization function creating an unmistakable warning through the artificial feel system and becoming active above 13.5°.

Two anti-stall systems are provided with No. 1 system having priority. The super-stabilization system applies down elevon deflection limited to 8°. This is done by the pitch auto-stabilization channel depending on angle of attack, pitch rate and airspeed declaration. At 19° a stick wobbler feature is introduced causing both control columns to pulsate by modulating pressure in the artificial feel jacks against any manual nose up stick force. The wobbler normally operates at 19° but has a phase advance of up to 3°. At speeds below 140 knots the super-stabilization commands a simultaneous 4° down elevon. 001 was used to develop this system and I had the opportunity to assess it when Jean Franchi brought 001 to Fairford for this purpose. I thought that the system showed the ingenuity of the French at its best and in my mind it afforded the protection provided by stick pushers, which are somewhat crude compared to this rather more delicate solution.

One deficiency in terms of meeting airworthiness requirements showed up in relation to longitudinal stability. Longitudinal stability is a complicated subject but it is sufficient to define it as the aircraft's tendency to return to its original attitude if disturbed from a trimmed state. In other words to reduce speed from a trimmed condition should require pull force by the pilot and a push force for increase of speed. These characteristics are laid down by the airworthiness (and military) authorities. Concorde showed itself to be neutrally stable up to M=0.95 when it became unstable up to M=1.3 when it returned to neutral stability. In practice this meant that there was practically no trim change up to M=2.0. It was therefore necessary to fit an automatic Mach trimmer similar to what happens on many aircraft in order to satisfy the requirements. This was a pity because the prototype was more pleasant to fly than the later models, which had the device fitted.

All aspects of handling performance engine, engine handling, relighting, noise abatement techniques and autopilot tests were covered jointly by 001 and 002. More and more induced failures were explored throughout the flight envelope.

FOURTEEN

Concorde: First Exposures

One of the most unusual features of the flight test programme was the exposure of Concorde to the public, potential customers and many VIPs. The appearance of both prototypes at the Paris Air Show in June 1969 has already been mentioned and caused enormous interest at the show and in the world press. Seeing these two great birds flying together must have been quite an experience, and was regarded as an excellent display, despite the fact that we had had no practice.

A few months later, 001 attained a speed of M=1.0 in October 1969, after which it was time to bring the airlines in. Four airline captains, James Andrew of BOAC (BA), Maurice Bernard of Air France, Vernon Laursen of TWA and Paul Roitsch of Pan Am flew 001 for about two hours each up to M=1.2 and 43,000 ft. Each flight covered various failure cases and a series of landings. This group concluded that the aircraft was easy to fly and that there were no problems foreseen in the training of airline crews, which is what we had hoped the feedback would be.

Farnborough followed in 1970 for 002 and in June 1971 001 made the first intercontinental flight of 2,500 miles in two hours and seven minutes from the Paris Air Show to Dakar. Further airline participation took place soon afterwards including Captain Alan Terrell from Qantas and Captain Red Steuben from Continental, both in 002. By now M=2.0 had been achieved, which enabled assessments to be made throughout the speed envelope.

Demonstration flights were given to Members of Parliament including the Rt. Hon. John Davies, Secretary of State for Trade and Industry and Lord Carrington, Secretary of State for Defence, as well as to the press. Although only very minor tests could be made on these flights there was always something to learn and they added to the overall experience. These flights were arranged in consultation with the Concorde Government officials.

In the April 1972, I took 002 to the Hanover Air Show with the Minister of Aviation, Michael Heseltine in attendance. We landed just behind the Russian Tu-144, which frightened the show organizers so much that neither of us was allowed to fly. The runway at Hanover was a bit short for the prototypes but quite adequate for Concorde. The Tu-144 scraped all its exhaust nozzles on landing, which apparently threw up a few sparks and got everyone watching unduly excited. I did not see it happen because Concorde was second to land some minutes later. When I was told by the show director that we could not fly, I was thoroughly cheesed off and told him so.

I spent two days looking over the Tu-144 with the 002 crew. The Tu-144 seemed to me to be a mixture of a modern supersonic transport and a Second World War bomber. The Russian crew spent a great deal of time in 002 copying a number of items in the cockpit, most of which we had decided to throw out on production aircraft. They were particularly interested in the engine display in 002 which was changed at airline request to a set of conventional engine instruments. Their co-pilot spoke perfect English with a strong American accent and was sadly killed in the Tu-144 at the Paris Air Show later on. He was a delightful person to meet, as indeed were all the Tu-144 crew. Our conversations were normally monitored by mysterious fellows in black mackintoshes who were invariably in attendance.

Far more startling than the Hanover Air Show was that I married Yvonne Edmondson, a widow with one daughter,

Sally. I had first met Yvonne at the home of Roger White-Smith, who had joined me at Fairford as General Manager. Roger's daughter Carole was a friend of Sally's; they had met at secretarial college in Cheltenham, an establishment where I understand neither of them enjoyed being. So, Sally was the link between Yvonne meeting Roger White-Smith and his wife, Jeannie. We married after a relatively short engagement at the church in the village of Ashley, near Tetbury, where Yvonne and Sally used to live. I had been a bachelor for forty-seven years and some adjustments on my part were clearly needed, but Yvonne's love and support, particularly when she hated aeroplanes, have always been a wonderful boost to my endeavours. We are quite unashamedly very happy and I owe her a very deep gratitude.

President Pompidou flew in 001 in May 1971 and followed this with a flight later in the year to meet President Nixon in the Azores. The use of Concorde by the President of France for overseas trips still continues today. Then followed a two-week tour of South America, which was trouble free and went very well.

Prince Philip and Prince Bernhard of the Netherlands, both expert pilots, flew 002 at M=2.0 during a test flight, although I did not let them attempt a landing. BOAC came in force at Chairman and Board level, along with Lord Rothschild and his 'think tank'.

HRH Princess Margaret and Prime Minister Edward Heath both had flights in 002 before we commenced a most ambitious tour to the Middle and Far East, Japan and Australia in June 1972. The tour was very heavily supported by maintenance and technical staff carried in an RAF VC10 and with a spare engine in a Belfast transport. This trip taught us a number of things which were totally unexpected.

The first leg was to Athens, where the take-off for Tehran was nothing short of frightening. The undulation and bumps on the runway caused violent oscillations on the flight deck. Johnnie Walker was in the left-hand seat and I was in the right. Our headsets came off and I was shouting at him to keep going because we were beyond the speed from which we could stop. On the way to Tehran, I said to Geoffrey Knight, Chairman of the Commercial Aircraft Division that I had 'a good mind to take the bloody thing back home . . .'. This feature remained with us for the rest of the trip especially at Bangkok and Singapore.

On arrival in Tehran, the brakes got hot at the end of the landing run so that, after coming to rest at dispersal, the thermal plugs in the tyres blew. The wheels had to be changed, which provoked some interest from an Iranian officer, who said that we would have to make an air test before the Shah could fly in the aircraft. I asked him who he was, whereupon he announced himself as His Imperial Majesty's personal body guard and that it was a rule that if any aircraft had work done on it an air test had to be made to his satisfaction before the Shah could fly. I told him bluntly, but politely, that we did not conduct air tests following a wheel change to which he replied that His Imperial Majesty would not be able to fly. This dialogue went on for some time between trips to his seat under a palm tree and to the aircraft. Each time he saw me, he first admired my Concorde pen, then my Concorde tie, then my Concorde cuff links, all of which were handed over very readily. Eventually he came back and said that he had decided that an air test would not be necessary.

We flew the Shah the next day up to M=2.0 over Iran and he had a spell in the left-hand seat, being a pilot himself. At this point, we learned something else. A large drop in outside air temperature occurred just as I was strapping the Shah into the left-hand seat. The aircraft was under autopilot control with Peter Baker in the right seat. The autopilot could not cope with the change and started a rapid climb followed by an equally rapid descent during which the airspeed exceeded the maximum limit by 20 k. We found out that the degree of temperature shears was near to prediction but that they took place over a much shorter horizontal distance than had been assumed. More will be said about the autopilot deficiency later.

The Shah was very favourably impressed and indicated at the subsequent meeting that Iran Air would purchase Concorde following normal negotiations. This was a great boost to the project, which was continuously under fire in many quarters. Iran Air eventually placed an order for three, although the deal was never finalized. The trip totalled seventy-three hours twenty-six minutes block time and sixty-two hours eighty-nine minutes flight time and included stops at Bahrain, Bombay, Bangkok, Singapore, Manila, Tokyo and Australia on the way out. On the way back stops included Dhahran, Beirut and Toulouse. The first landing at Bombay was quite an experience because an enormous crowd had assembled to watch Concorde land. Most of the onlookers came from the Air India maintenance base, as well as airport staff. Unfortunately, the crowd control had broken down and the onlookers were packed along each side of the runway so it was rather like landing on something that resembled a football pitch. The other surprise was finding a sacred oxen in the middle of the terminal building. (The full itinerary is shown as Note 1 at the end of this chapter.)

Apart from the fuselage response to rough and undulating runways, two other main lessons were learnt. Firstly, the Met man was incorrect in suggesting that Concorde would encounter cloudless skies at its operating altitudes. A major tropical storm was present in the Bay of Bengal between Bombay and Bangkok and Concorde 002 was in cloud at 60,000 ft. Secondly, the sudden temperature shears experienced highlighted the inadequacy of the autopilot to keep the aircraft within normal speed limits.

For the first part of the tour with Geoffrey Knight and Sir Geoffrey Tuttle as hosts, Mr Michael Heseltine and his wife were on board. At Singapore they were replaced by Lord and

Lady Jellicoe and Sir George Edwards also joined the party. Lord Jellicoe and Michael Heseltine were representing the British Government. I had not met the Jellicoes before, although he was another old Wykehamist, and they were tremendous fun and we used to meet after we returned home. George and Philippa Jellicoe were so supportive and it was a tremendous pleasure to have them as well as the Heseltines with us.

Between Bombay and Bangkok Mrs Heseltine became extremely ill but fortunately we had a BOAC steward, Fred Clauson, on board who provided the necessary treatment as best he could in the rather primitive passenger accommodation on the prototypes. When we reached Singapore, Mrs Heseltine told me she had eaten shellfish and mangoes in Bombay. I did mention that I thought people who eat shellfish and mangoes in Bombay are likely to become very ill! Fortunately, by the time we reached Singapore, this delightful and charming lady had recovered.

The reception in Tokyo was somewhat mixed because of Concorde's noise and smoke. We landed over the sea and took off in the opposite direction at Haneda in order to minimize these effects. The British Ambassador and his wife, Sir John and Lady Pilcher, flew out with us to Darwin as Sir John's tour of duty in Tokyo ended at that time of our departure.

The next bit of excitement was a bomb scare in Darwin. I was woken in the middle of the night to be told that a bomb was reported to be on board. As the flight to Sydney was scheduled for an early departure, I left the search operation to others and went back to sleep. Like many bomb scares it proved to be false, but such warnings cannot be ignored. The telephone caller who gave the warning was never traced to my knowledge and remains anonymous.

The prototypes were fitted with a tail bumper similar to naval aircraft and it often touched the runway on landing. This happened at Sydney and threw up some sparks, which got the fire brigade tremendously excited. Australia's reaction to Concorde was split over noise, but many of those in the right places were very supportive and the crowds on arrival at Sydney were all of 20,000. Mention of a supersonic route was made on several occasions, which did actually come to fruition and was flown during the route proving trials.

On the way home, the sector between Bangkok and Bombay posed a difficulty in the sense that any decision to divert to the nearest suitable airport, Delhi, had to be made before descent below 25,000 ft. At the point of decision, the weather was reported to be very good so I decided to continue, but a few minutes later I could see some rather nasty looking cloud ahead. The monsoon had arrived and was over Bombay, and we went into cloud and broke out at about 400 ft in very heavy rain. Thanks to the reverse thrust and tail parachute, I had no trouble stopping, which is more than I can say for the Boeing 707 that landed an hour later and went off the end of the runway. This was enough for me, so I cancelled the demonstration flight on the grounds of a 'technical' snag, the only loss of a scheduled flight during the whole tour.

The anti-Concorde brigade were well organized for our arrival at Heathrow. Postcards registering complaints had been specially prepared, but their impact was minimized when it was seen that many of them had been posted a day early.

NOTE 1 – THE FULL ITINERARY:

2 June Fairford–Athens
2 June Athens–Tehran
4 June Demo (2)
6 June Tehran–Bahrain
6 June Bahrain–Bombay
7 June Bombay–Bangkok

7 June Bangkok–Singapore
9 June Demo
9 June Demo
11 June Singapore–Manila
12 June Manila–Tokyo
13 June Demo
14 June Demo
15 June Tokyo–Manila
15 June Manila–Darwin
17 June Darwin–Sydney
20 June Demo
20 June Sydney–Melbourne
21 June Demo–Melbourne–Darwin
23 June Darwin–Singapore (one hr twelve min at M=2.0)
24 June Demo
25 June Singapore—Bangkok
25 June   Bangkok–Bombay
28 June Bombay–Dhahran
29 June Demo
30 June Dhahran–Beirut
30 June Demo
30 June Beirut–Toulouse
1 July Toulouse–London Heathrow
5 July Heathrow–Fairford

FIFTEEN

Concorde: Further Testing

The third Concorde (pre-production No. 1) joined the test programme on 17 December 1971, when I flew 01 from Filton to Fairford. The main snag was with the landing gear which started cycling up and down following first retraction, but a quick selection down stopped this nonsense but curtailed the flight. 01 was not fully representative of the production aircraft in many areas, although it did have the revised nose and visor, but not the rear fuselage extension. I felt that the revised visor was a big step forward and it really transformed Concorde flight operation. The differences were extensive which meant that we had to repeat in-flight flutter tests in the same manner as I have described earlier for the prototype and the same methods of forced excitation of the structure again proved to be satisfactory.

For nine months 01 flew without an intake control system, which was supposed to be representative of production, and this meant that speed was restricted to M=1.5 max. A lot of handling and performance checks were made during this period. Water ingestion trials were also conducted at Toulouse as the water trough was better than the one which we had installed on the south taxiway at Fairford.

In August 1972, 01 went back to Filton for an up-date and reappeared on 15 March 1973 after extensive modifications including the air intake control system. 002 continued with its allotted programme of handling and performance tests. Take-off and landing techniques were developed at this time which introduced the procedure of closing the throttles at 15 ft prior to landing. This ensured much greater consistency in measured distances than had been achieved previously.

Another SBAC show at Farnborough and special demonstration flights over the UK took place in September with another weather diversion to London Heathrow on one of the show days. A landing considerably overweight brought a lot of noise complaints, as the technique of decelerating approaches was not in use at this stage. Later a procedure of decelerating from 210 knots at 800 ft to threshold speed was developed. British Airways modified this to be 190 knots at 800 ft. Continental Airlines came to Fairford for a two-flight assessment which went very well.

Demonstration of satisfactory de-icing capability has always been quite a difficult aspect of certification because getting the right amount of ice to stick on an aircraft can be very time consuming. A new method of flying behind a water tanker was beginning to emerge and some flying was done behind a USAF KC-135 and behind a Canberra operated by Boscombe Down. This enabled the aircraft under test to ice up half the airframe wing and engine installation. The Canberra tanker had some teething problems with the water release valve, so that on one occasion I saw something like a football flash past the cockpit. After this I was not so keen on flying with the Canberra tanker.

At the beginning of 1973 I went back to Athens in 001 with André Turcat for him to get first-hand experience of the runway response problem. After the first take-off I said to him 'There you are, I think that it is quite unacceptable'. To which he replied, 'I would not say that it was unacceptable, but I would agree that it might interfere with emergency procedures' and, after all, the landing gear was a French responsibility. We spent two days in Athens carrying out further take-offs before flying back to Toulouse via

Marseilles. It was very foggy at Toulouse, so we made an automatic landing.

I had no sooner finished this trip and I was off again to South Africa for high-altitude performance trials at Johannesburg (5,600 ft above sea level). We routed out via Las Palmas, Robertsfield in Monrovia and Luanda. It had been decided that on landing in a particular country we would fly that country's flag and the Union Jack wherever Concorde arrived. This meant opening the two side windows of the cockpit to push the flag out, so we only did it after landing, but not before take-off.

On landing at Luanda, the flags were pushed out and I was soon asked what flag we were flying, to which I very proudly replied 'The National Flag', not knowing that the flag we were flying was in fact the Rebel flag. I had some difficulty in convincing the authorities that it was a genuine mistake. Thankfully, I won in the end and was presented with the correct flag, so jail was avoided. However, as we taxied out for take-off, as per normal procedure, no flags were flown. This unfortunate apparent lack of protocol caused great offence as 002 was suddenly surrounded by soldiers with rifles and machine-guns. I taxied back to the apron where our representative was standing with a British passport in one hand and two fingers up on the other, which was supposed to tell me to fly the Union Jack one side and the new flag on the other. With that I closed down the engines and disembarked, the offence being that I had not flown the proper flag taxiing out for take-off. Jail became a strong possibility and was only narrowly avoided; I made sure that there was no similar mistake on the return journey.

A large crowd welcomed us at Johannesburg but I was a bit put out when I noticed that the 'loo' cleaners wore orange overalls very similar to our flying suits. Our reception was more impressively heralded by the large crowd than when I had arrived in the Vanguard some years before, but even in those days new aircraft types tended to provoke great interest.

The test programme at Johannesburg was mostly measured take-offs and measured landings with some two-engine climbs. Various abuse cases (speed too low or too high) were covered and a critical test of landing at target threshold (VTT) minus 5 knots on the forward centre of gravity was so easy on Concorde compared to all the other aircraft on which I had previously made a similar test so that I also made one at VTT-10 K. The FTG test pilot, Gordon Corps of the CAA, took part in some of the flying. While we were in South Africa, demonstration flights were carried out at Cape Town for many VIPs, including the South African Prime Minister.

The Concorde had no water ballast system for increasing weight. Consequently, the required take-off weight was achieved by the amount of fuel loaded. After each heavy weight take-off, fuel had to be jettisoned to get down to a maximum landing weight, so that a further measured take-off test could take place.

The return flight to Fairford was the same as the outbound except that a visit was made to Kinshasa from Luanda.

As soon as 01 came back from Filton, the flight envelope was rapidly expanded so that the major task of proving the air intake control system could commence. Engine handling was explored up to M=2.165 and engine surge tests were deliberately introduced by inching the ramps manually. Interactive surges with the adjacent engine occurred on many occasions. Pushovers to 0g (weightless condition) and application of sideslip were part of the routine. The majority of the testing was done on engines 1 and 2. The surging of No. 1 engine introduced smoke through the air-conditioning system into the cockpit, causing several members of the flight crew to develop sore throats. Fortuitously No. 4 engine proved to be the most critical later on, so testing was switched mainly to Nos 3 and 4 engines. A local vacuum-cleaning firm in Cirencester had to be brought in to clean out the air-conditioning system.

Another significant factor arose in that the critical low temperature cases could not be reached flying out of Fairford. The low temperature required occurred near the Equator and it was too far south to be attainable from Fairford. Consequently, 01 went to Tangier early in 1974 and carried out a major programme on the intake control system. We had purchased a small computer for on-the-spot analysis, and this was installed in a furniture van bought locally second hand which travelled out to Tangier by road and sea ferry.

John Cochrane was the spearhead of these tests and did a wonderful job. During the tests 01 reached M=2.23. On one flight all four engines surged which prompted John to say, 'I thought that I had been sent for . . .'.

The intake control laws were changed several times and this was done in a tent alongside the aircraft. Bob McKinlay also played a leading part in the programme. More intake testing in the UK followed a second trip to Tangier. After this, 01 appeared at the SBAC Show in 1974 and then went into the de-icing programme using the Canberra tanker. FAA requirements were quite clear that flight in actual icing conditions had to be demonstrated. John Cochrane took 01 to Moses Lake, Washington via Bangor, Maine for this purpose and after some difficulty managed to get sufficient ice – up to 2 in – to stick and so satisfy the requirement.

Meanwhile 002 had carried out a series of environmental flights from Prestwick before I went to Madrid for high-temperature trials based at Torrejon, which consisted mainly of take-off and climb measurements. Having spent a lot of time trying to find natural ice in the UK, eventually the work was transferred from 002 to 01. This was the last major task on 002 except for demonstration flights including one for Her Royal Highness, the Princess Anne. The last demonstration at Weston-super-Mare Air Show nearly ended in disaster when, on returning to Fairford, Eddie McNamara selected the landing gear down while in a 45° turn. There were two loud bangs as the main retraction jack and the main side stay broke away from the left leg. This was confirmed by visual inspection, which meant that the left landing gear which was not showing a green locked down light was unsupported. John Cochrane took over control and prepared the crew for an emergency landing, which he skilfully executed by making a very smooth touch down and holding weight off the left gear by using right roll elevon. After coming to rest, jacks were put underneath the aircraft to prevent any landing gear collapse.

By 1974, the flight test programme was in full swing and the No. 2 production aircraft joined the programme in February when I flew 202 from Filton to Fairford, reaching M=1.43 on the first flight. I was concerned by what appeared to be airframe buffet and actually sought André Turcat's opinion by going to Toulouse for comparison with 201. We concluded that there was a slight difference to the prototype but it was otherwise acceptable. Measured take-offs, measured landings, various climb configurations and cruise performance were the order of the day. Engine handling and relighting were also high on the list and initial assessments of noise abatement procedures for New York, Kennedy were made.

Engine intake handling was checked by flying well south to reach cold temperatures and then landing at Casablanca. I made two flights with Aerospatiale firstly from Paris to Rio via Dakar in a block time of six hours thirty-nine minutes and six hours thirty-five minutes on the return with a turnaround in Dakar of thirty-seven minutes. I also went Paris–Boston–Miami–Boston–Paris in Concorde 02.

The two prototypes had really come to the end of their programmes and 001 went to the French Air Museum while 002 remained at Fairford following its landing gear incident pending disposal.

In July 1974, Prime Minister Harold Wilson and French President Giscard d'Estaing agreed on an initial batch of sixteen Concordes. This of course only covered the BA and Air France orders and four white tails.

I was sitting in my office one morning when the telephone rang and a voice said, 'It's Tony Benn'. I saluted the telephone as the Minister continued, 'I hear that you are off to Bahrain next week, so I thought that I would come and have a flight with you and bring some of the chaps!' I asked who the chaps might be and was told, 'The Shop Stewards from Filton, Weybridge and Hurn of course'. I was asked about seating and to cut a long story short, we settled on forty. The exercise took place on a Saturday afternoon following lunch for all involved. I was a bit surprised that all the stewards greeted the Minister as 'Tone', but then life is full of surprises. A speech followed the flight and I was most impressed when the senior steward thanked 'Tone' and 'Mr Trubshaw' for the flight, I thought 'Fame at last . . .'.

Two days later and the day before I was scheduled to leave for a long programme of hot-weather trials in Bahrain and runway response tests in Singapore, I flew from Fairford to Weybridge in our small twin-engined Beagle 206S communication aircraft to see Geoffrey Knight. On the way back, I had what could have been a big drama. Soon after take-off from Brooklands I experienced extreme rough running of the right engine, so I decided to stop it. I cut off the fuel supply and feathered the propeller whereupon the vibration stopped. I had a look round, pleased with my handiwork, only to find that the right-hand propeller had gone. I decided on an emergency landing at Farnborough, which was nearby and their ATC had been informed of my flight through their airspace by the fire station at Brooklands, so they were not surprised when I called them on my radio. The controller answered immediately and obviously recognized my voice because he said, 'Haven't seen you for a while', to which I replied, 'You will see me any minute because I've lost my right-hand propeller!' I made a safe single-engine landing at Farnborough and was greatly relieved that the propeller had been found immediately and conveniently in Brookwood Cemetery near Woking. I finished my journey by car late in the day. I was extremely lucky that it was the right-hand propeller that came off because the direction of rotation meant that it fell away from the aircraft. If it had been the left one, the story might have been quite different as it would have probably entered the cockpit. No fuss was made of the incident – my secretary, Caroline Paginton, telephoned my wife and said I had been delayed at Farnborough and would be a bit late home.

Sir Geoffrey Tuttle, who had become a great friend and supporter, came with us to Bahrain and Singapore as he loved these overseas trips. The outbound flight to Bahrain was made over Russian air space via Tehran. In the course of the trials at Bahrain, demonstration flights to Kuwait, Doha, Dubai, Muscat and Abu Dhabi were included in the schedule. The flights from Bahrain were all performance related, climbs and cruise and systems performance under high-temperature conditions. Bahrain was an ideal location as there were no restrictions being on the edge of water. The flight normally went eastward towards Pakistan, turned and came back westward to Bahrain. Our deceleration point on the return leg was abeam Dubai, at about 180 miles from Bahrain. My wife, who was staying in Bahrain, kept saying that she had heard us boom. This could not be the case, or so we thought at the time, but what she was in fact hearing was secondary boom, about which we knew practically nothing.

Being in Bahrain necessitated courtesy visits to Kuwait, Doha, Dubai, Abu Dhabi and Muscat, all of which I enjoyed tremendously finding the Gulf area quite fascinating. When we were not flying, the Ruler of Bahrain graciously allowed us to use his private beach. I then went on to Singapore to assess the runway response characteristics of the production aircraft. The answer was just as unacceptable as on the prototype not only in my opinion, but shared by Gilbert Defer of Aerospatiale and Gordon Corps of the CAA. In due course the Singapore runway was re-laid at the expense of HMG. Concorde take-offs had to be approved for individual runways when airline service was started pending the retrospective modification of the landing gear, which incorporated a two-stage oleo in its modified form. The runway tests at Singapore included abuse cases, e.g. early rotation as well as normal and late rotation. The return to Fairford was made via Bahrain in seven hours thirty minutes flight time and seven hours fifty-two minutes block time, including the stop at Bahrain.

Performance testing and further development of noise-abatement procedures were the main tasks on 202 at this stage. This led to another overseas assignment to Casablanca, where the New York Kennedy noise monitoring points were laid out in the adjoining desert to simulate take-offs on 31 left at Kennedy. Our pattern required a low-level turn not above 100 ft to the right whereas Kennedy requires a left turn for the Canarsie beacon when departing from Kennedy 31 left. Anyway, the result was the same.

After each take-off, climb and cruise performance measurements continued including two-engine and three-engine cases in day and night conditions. The low level turn in the noise-abatement procedure needed a bit of selling, so British Airways and CAA took part and everyone agreed that the precise handling qualities of Concorde made this manoeuvre easy to fly and perfectly acceptable even in the case of engine failure in the turn and pushover cases to 0.75 g.

Apart from a fly past at Rabat and Marrakesh, we made one demonstration flight to local VIPs but one turned out to be a member of the PLO (Palestine Liberation Organization). Frantic telephone calls by John Ferguson-Smith, our managing director who had flown out specially for the demonstration flight, to the Foreign Office and so on resulted in a decision to leave matters as they were but it was agreed that John Ferguson-Smith would stay close to him throughout the flight. As luck would have it, No. 3 engine failed causing a massive surge when we were accelerating through M=1.7 just as they were walking towards the flight deck. Both returned rather hastily to their seats, but with their linen out of order.

At the end of the trial, I flew by Pan Am to New York for a meeting with the Port of New York Authority on the matter of noise where I met up with Mick Wilde, the BAC Concorde Project Director at that time. Although we were received in a most cordial manner and got on very well in presenting the measured noise levels and the technique approved by the Anglo-French authorities, it was merely the start of a long and bitterly contested struggle over the next two years.

After this I went back to Casablanca for another week. Confidence in Concorde had now reached a level when overseas visits were routine but necessary to cover all the numerous conditions needed to satisfy the airworthiness requirements. 01 proceeded to Nairobi for tropical icing tests, which were successfully completed.

This was the last major test task for 01, which started the BA crew training in March 1975 prior to the endurance flying programme on the aircraft 204, which I first flew from Filton at the end of February 1975. 204 had to complete an extensive test schedule covering the whole flight envelope including handling, performance, engine handling, relighting, electrical, pressurization and cold soak. Final flight training for BA was undertaken by using both 202 and 204. A special category Certificate of Airworthiness was granted for aircraft 204 at the end of June 1975, so that route flying could start.

It is as well to record some of the special aspects of

operating Concorde because all this had to be conveyed to the nucleus group of BA pilots, consisting of Captains Mickey Miles (Flight Manager), Brian Calvert (Deputy Flight Manager), Norman Todd, Pat Allen, Pete Duffy, Tony Meadows (Training Captains), with Captains Chris Morley and John Eames to follow, all of whom were designated to take part in the route flying, which was part of certification.

For taxiing, care has to be exercised when turning sharp corners or making tight turns on the ground as the pilot sits 38 ft ahead of the nosewheel and 97 ft ahead of the main wheels. This necessitated putting the nose over the grass during part of the turn. The fuselage tends to be a bit 'lively' especially if any flats have developed on the tyres due to lengthy standing at heavy weight. Only idle power is necessary to keep rolling and the carbon brakes are very powerful.

It is not possible to hold full power against the brakes. Reheat is pre-selected, brakes released and full throttle applied with time clocks started. The acceleration is rapid and reheat light up 'go' lights are checked at 100 knots. Rudder pedal steering is used to keep straight. The bugged decision speed (V1) is reached, followed quickly by the bugged rotation speed (Vr). This requires a definite movement of the control column to apply up elevon. Concorde is peculiar in that the wing does not produce lift until the aircraft is rotated. The pitch attitude required is bugged on the attitude indicator and is usually about 13°/14°. This attitude is held as the aircraft continues to accelerate after lift-off and further rotation to about 17° or 18° is needed to hold the required speed of 250 k; the maximum speed permitted over United States airspace for flight below 10,000 ft is 250 k. At the noise abatement cut-back time, the reheat is switched off and the throttles reduced to the pre-determined power setting while still maintaining 250 knots at reduced attitude. At some locations like Bahrain, noise abatement is not required in which case speed can be allowed to increase to 380 k at full throttle but in a climb mode instead of take-off mode by moving one switch. Once the nose and visor are raised the flight deck is uncannily quiet. Normally the climb is restricted to below 28,000 ft at M=0.95 until acceleration to supersonic speed can be permitted, which happens as soon as the aircraft is over the sea.

The climb then continues with speed increasing to 530 k and M=2.0. At this point a cruise climb technique is adopted and the aircraft drifts up towards up to 60,000 ft, the absolute cruise altitude limit because of air-conditioning requirements. Structural temperature is fed into the high-speed warning system in addition to the traditional Mach number and air speed. It is a fairly frequent occurrence that outside air temperatures on the North Atlantic are higher than standard which may result in the indicated cruising speed being reduced below M=2.0. The temperature and its relation to standard are displayed to the pilot. Reheat is used between M=0.95 and M=1.7 for the acceleration.

Autopilot control is normal practice but Concorde is not difficult to fly manually, as its control is so precise. Engine failures do not pose a problem except supersonic flight is not maintained on three engines. One peculiar feature is that engine failure supersonic causes the aircraft to roll gently the wrong way due to lift from the opening spill door in the affected intake until sideslip takes over and produces roll response conventionally, but it is all very innocuous. Nevertheless, Concorde travels at over 20 miles per minute during cruise which is slightly faster than a rifle bullet, so everything happens quickly. When flying on an airway Concorde operates like any other aircraft but over the sea it uses its three inertial navigation systems, which are tied into the autopilot to travel from one Way point to the next. The inertial navigation system can be updated from VOR/DME stations and this results in an incredibly accurate system. Throughout the flight, the flight engineer is controlling the fuel system in order to ensure the correct centre of gravity, so that between 0° and 1° down elevon is achieved in cruise. The deceleration and descent are planned to be below M=1.0 approximately 30–5 miles from the coast. Initial deceleration to M=1.6 is made in level flight, whereupon the throttles are retarded further and descent begins at 350 k. Descent is maintained to the assigned altitude, normally 39,000 ft on the way to New York. Thereafter, further descent clearance is obtained, usually in steps and the aircraft is duly positioned for approach and landing, commencing at 190 knots. The landing gear is lowered on intercepting the glide path and the nose/visor which has been set to 5° and visor down, is then lowered to 12½° down for landing.

It is normal practice to use automatic throttle control for approach and landing. Manual throttle is regarded as an 'abnormal' procedure which is taught in training but Concorde, being a delta, is on what is known as the 'back side' of the drag curve on approach. Use of auto-throttle is more accurate and reduces crew workload. On many aircraft, use of automatic throttle on a manually flown approach results in instability but this is not the case with Concorde due to the auto-throttle law, which includes an acceleration term. The attitude on the approach reaches about 11° nose up as speed is reduced to the target threshold speed which is typically about 160 knots. This attitude gives some pilots a feeling that they are going to land half way down the runway as mentioned earlier but is trained out very easily.

For minimum noise approach speed is maintained at 190 k until 800 ft, when target threshold speed is selected. Height calls are made by the flight engineer (or the co-pilot) below 400 ft and particularly 100, 50, 40, 30, 15 ft. At 15 ft the throttles are closed and backward movement of the control column is required to counter a nose down pitch tendency due to ground effect and throttle closure, so that a constant attitude is maintained to touch down. At this point the pilot's eyes are still 38 ft above the runway. The ground effect, which is caused by the wing squeezing the air between itself and the ground into a sort of cushion can be readily heard. After touch down the nose is lowered on to the runway, reverse thrust is applied and wheel brakes are used as necessary. Reverse is cancelled below 75 k in order to avoid re-ingestion of engine exhaust gases. It all sounds too easy. All the abnormal and emergency procedures are all covered during initial and recurrent training either on the simulator or in flight.

SIXTEEN

Concorde: Route Proving, Flying, Training and Entry into Service

Training the nucleus group was very straightforward because nearly all of them were very good pilots and thoroughly sharp. Mickey Miles was not in full flying practice at the start which made conversion more difficult for him. With the training complete, the first route proving started on 7 July 1975. I positioned to Bahrain with Pat Allen to operate the return flight Bahrain–London. The British side of the route flying covered the Middle East, Far East, Australia and Gander. The French mostly used routes to South America and Gander.

The initial arrangements were for the company test pilot to be in command for phase one which was the equivalent of route training but for the airline crews to take over for the second phase. There was, however, a change of heart by the airworthiness authorities who laid down on behalf of the Governments that a company test pilot had to be in command all the time, albeit occupying the jump seat. This went down very badly with BA and I got a telephone call at home from the General Manager Flight Training, Captain Phil Brentnall saying that this arrangement was '. . . unlikely to be acceptable to British Airways . . .'. I told him that it was not my decision and that he was stuck with it, and eventually the flurry subsided and we were able to get on with the job. We normally carried a full load of passengers who were looked after by cabin crews drawn from Air India, Singapore Airlines and Gulf Air because there was a dispute with BA cabin staff over Concorde rates and conditions.

Although the route proving was hard work, it was extremely enjoyable. There were a fair number of snags, particularly with the engines where the 'vaporisers' in the combustion chamber came adrift, which necessitated an engine change. One of these happened when the Australian Prime Minister, Mr Gough Whitlam flew from Melbourne to Singapore. His presence had attracted the attention of Mr Gerald Kaufman, MP, who was livid when I told him that the flight was terminating at Singapore until the engine was changed. I had to explain to him that there was nothing I could do about it even if I had the Lord Almighty on board. In the end I left him in the hands of Sir Geoffrey Tuttle.

One week was spent flying between Singapore and Melbourne with a supersonic route across Australia running just west of Alice Springs. The Minister of Aviation had worked very hard to get approval and was extremely disappointed when the route was not followed up in airline service. The plan fell apart due to the demands of Qantas, who expected additional B-747 flights from Sydney to London, which would have resulted in an overall disadvantage to BA.

The second phase put two BA captains in the pilot's seat with the BAC pilot on the jump seat in command. I had a minor drama landing at Bombay one night. When the reverse thrust is selected on landing, the control column must be held fully forward in order to prevent the nose rising. On this occasion, this did not happen, the nose came up which scratched one of the engine exhaust nozzles and put a hole in it. I was concerned that sudden cancellation of reverse would result in the nose leg banging down, so I was shouting instructions to Mickey Miles and Pat Allen. All ended well except for the hole in the exhaust nozzle. I got a fitter from Air India to put a patch over it and we continued on to Singapore at M=2.0.

Bob McKinlay was a bit shattered when he saw the standard of repair after our arrival.

Many interesting people flew as passengers including the Archbishop of Canterbury, Lord Coggan who said after his flight to Gander and back to London that he had '. . . never been closer to God'. I was pleased with the way that the route flying went and I was quite sad when it was over. Flying Concorde from A to B really gets the message home of a shrinking world in a supersonic transport.

The bulk of the BAC commitment was shared between John Cochrane, Roy Radford and myself. Peter Baker refused to do any crew base training as he felt that all the training should have been done by our ex-BAC1-11 training captains, John Hackett, Paddy Cormican and Alan Smith but it had not been possible to give them enough Concorde exposure. They did, however, do a great deal of work on the training simulator, which originally belonged to the Royal Bank of Scotland, and prepared the training briefs for actual flying training. I, therefore, left Peter Baker out of the route flying, but somebody had to be left to carry on at Fairford and so he had a ball.

Most of the simulator commissioning work, which was very exhaustive and took a long time, was in the hands of Johnnie Walker. The end product was a good representation of Concorde handling characteristics, but the visual display of the runway etc. was not good enough in its original form for practical landings. These all had to be done on an aircraft and covered all the abnormal cases such as no auto-throttle, no height calls, visor up etc.

At the end of the route flying some reorganization took place at BA and Brian Calvert took over as Flight Manager. In October and December 1975 the French and British Civil Aviation Authority respectively granted Concorde a Certificate of Airworthiness. British Airways formally objected to the granting of a Certificate of Airworthiness because of the tendency of Concorde to overspeed due to the temperature shear not being fixed, although a solution had been tried on the French aircraft 201 operating from Kuala Lumpur in October. I joined up with Jean Franchi and Gilbert Defer to define a change to the autopilot and auto-throttle systems which armed the auto-throttles in cruise so that the auto-throttles could intervene if Mach number increased above 2.0 by reducing engine power. It was a very elegant modification, typical of French ingenuity. The officials' test pilots had witnessed the results but it was then a question of retro-fit to the in-service aircraft.

I attended the ARB Council Meeting which met to discuss the granting of the Certificate of Airworthiness and BA were represented by Brian Calvert. The BA objection was overruled and the Certificate granted but BA had painted the picture so black that the ARB decided that a company test pilot had to be on board for an unspecified period when commercial service started. This went down very badly within BA, as personal egos were at stake among other considerations. So John Cochrane and myself were appointed CAA Flight Inspectors and one of us flew on the first seven flights incognito during the first month of BA airline service. Our joint report satisfied the CAA who removed the requirement on our recommendation.

Commercial service started on 21 January 1976 when British Airways flew to Bahrain and Air France to Rio de Janeiro with simultaneous take-offs from London and Paris. BA used their first delivery, the sixth production Concorde 206 (GBOAA) which I first flew in November 1975 and which was delivered to BA following completion of the production test schedule. For the flight to Bahrain I crept on board with the passengers but turned left at the top of the stairs into the cockpit and sat down on the jump seat. Norman Todd was in command with Brian Calvert as co-pilot on the way out and the other way round for the return flight the next day. John Liddiard was the flight engineer on both flights. Mrs Thatcher (Lady Thatcher) came to see the flight off, while on-board guests included Sir George Edwards, Lord Boyd-Carpenter, Chairman of the CAA, the Duke of Kent, Peter Shore, Secretary of State for Trade and Eric Varley, Secretary of State for Industry and Sir Leonard Cheshire VC.

The route to Bahrain involved flying subsonic to Venice, then south down the Adriatic reaching M=2.0 and turning left to fly south of Crete and Cyprus, north of Beirut over Syrian airspace and Saudi Arabia and on to Bahrain, where the ruler gave a fabulous banquet for all the crew and passengers. I was so incognito that I was nearly left out but some kind person, Norman Todd I believe, put this right. This was indeed a historic day marking the beginning of supersonic travel.

BA had to manage with one aircraft until mid-February when 204 (GBOAC) was delivered after refurbishment following the route flying. Meanwhile, on 4 February the enormous efforts by the Governments, BA, Air France and the manufacturers to obtain Concorde services into the USA were rewarded. US Secretary of Transportation, William T. Coleman approved BA and Air France to operate two services each per day to New York and one service per day to Washington on a trial basis for sixteen months. But it did not end there. The implementation of the Coleman decision for Dulles, Washington was not too difficult, because the airport is FAA operated and the FAA are linked to the Department of Transport, although some legal objections were raised but were easily overruled. But this was not the case at New York, run by the Port of New York and New Jersey Authority, and the Coleman decision raised the sensitive issue of the rights of the federal government against the rights of states. Noisy aircraft operating in and out of Kennedy were being bitterly opposed by the 'locals' and the prospect of Concorde operations was too much.

In March 1976 the Port of New York and New Jersey Authority banned Concorde flights in and out of Kennedy airport. The airlines filed a suit against them, challenging the Authority's right to do this. The PNYA responded with a requirement to conduct a six-month study of Concorde noise at Washington, London and Paris. Other legal cases helped to occupy most of 1977 as well. Judge Milton Pollack of the Federal District Court gave rulings in favour of the ban being lifted because it was illegal, saying that the delay in the Authority setting noise standards for Concorde was 'discriminatory, arbitrary and unreasonable'. He said that the aircraft had not been given the opportunity to show that it was environmentally acceptable. A further appeal by the Authority was rejected in the US Supreme Court, who stood by the ruling given by the Federal District Court. This ended the ban but it was not until November 1977 that BA and Air France commenced scheduled flights into and out of Kennedy airport, New York.

In the meantime BA and Air France had commenced commercial operations to Dulles airport, Washington on 24 May 1976. The departure times from London and Paris were adjusted to enable the two aircraft to land in quick succession. This was indeed achieved with BA landing first and the taxi to the terminal was well rehearsed so that both aircraft arrived on the ramp, stopped and raised their noses and visors in a completely synchronized manner. A special descent technique had to be devised in order for Concordes to comply with the height requirements of danger areas on the inbound route and at the same time meet the Mach=1.0 point. This became known as 'the fall of the wall', which involved descending from 52,000 to 43,000 ft at a constant Mach no. of 1.3 with a pre-set altitude capture of 39,000 ft selected. At altitude capture speed was and is reduced to M=0.95.

A lot of work was done by Bob McKinlay and Henri Perrier in preparing a detailed presentation of how Concorde operations at New York would work in practice. This study showed that most departures could use runway 31 left or 22 right. It also showed the actual noise pattern was similar to the Boeing 707-320B and that Kennedy airport's noise criteria would be met, especially as the normal take-off weights would be approximately 14 tonnes below maximum. The excellent and precise handling qualities of Concorde played an important part because of its exceptional ability to operate in crosswinds and tailwinds, its high level of performance and its ability to fly tighter patterns than most conventional aircraft.

All these facts were compiled in a massive book presented to the Port Authority. It represented the most detailed analysis and study of aircraft noise impact ever made. The determination of all those who were involved in winning the 'battle' has to be admired very greatly.

The years of frustration were soon to be over and, once all the pressure groups had finished, operations into the USA became a normal daily occurrence. But like so many trials and tribulations that Concorde has had to face, there was a harsh penalty to be paid. The inability to gain permission to land in New York played a major role in airlines cancelling their options.

The same routine of arriving together was adopted for the first arrivals at New York on 22 November 1977. The flights were preceded by a short trial by aircraft 201 over a period of three days mainly to check and clear the use of two runways at Kennedy, which were known to be rough and undulating. Jean Franchi commanded the aircraft, accompanied by the official test pilots and a number of airline captains. Actual noise measurements were naturally made and were all well within limits. Enormous press coverage had portrayed a grotesque scene but it was nothing like what had been brandished around and many had to admit that it was a non-event.

SEVENTEEN

Concorde: the Final Stages

In 1976, also an eventful year, I took 002 to its final resting place at the Fleet Air Arm Museum at Yeovilton. Following the landing gear incident on its previous flight, the landing gear was locked down. I carried out a short air test first at Fairford and I thought how much nicer to fly 002 was, with its marginal longitudinal stability, than the production aircraft which had been fitted with artificial devices to show compliance with the airworthiness requirements.

I needed good weather at Yeovilton because the runway length there was such that a low fuel load had to be carried to keep the landing weight as low as possible. Various telephone calls confirmed good weather, but on arrival I realized that we had been given a bum steer and the weather was low cloud and poor visibility requiring a precision approach by radar.

The Yeovilton controllers were not used to the relatively high approach speed of Concorde (about 150 k in this case) so I had to make two attempts at landing and had insufficient fuel to go back to Fairford or Filton. On the second approach John Cochrane and I saw the runway in time to execute a landing. I was put off a bit by the Commander Air, standing on the runway waving his bats as per a carrier landing and hoped that he was going to get out of the way, as I was determined to land. He obviously did but I thought that I might have killed him until I saw his smiling face in the reception party, which included Prime

Minister Harold Wilson, Mr Gerald Kaufman MP and Captain Jimmy Abraham RN, who had been with us at Wisley on the Scimitar Programme. And so I said *au revoir* to a very important friend, Concorde 002 (GBSST). Years later the Museum built a hangar for 002, where she still sits with the BAC 221 and HP115 beside her.

The next major task involved testing a higher take-off thrust rating known as 'contingency' which Rolls-Royce were able to offer for use in emergency. This was done by over-speeding the engine by 2 per cent N2 (high-speed compressor speed). It may not sound much, but it involved another trip to South Africa. Aircraft 202 was used for this and I made an unofficial record flight from Fairford to Cape Town via Robertsfield where the refuelling stop lasted forty-five minutes. The actual flying times were three hours twenty-six minutes to Robertsfield and three hours seven minutes to Cape Town. No publicity could be given because Monrovia was a Black country and did not wish to be associated with South Africa. It was on this basis that we were allowed to use Robertsfield. The test programme was mostly take-offs with some extra performance measurements and auto-lands thrown in to use up the fuel out of Johannesburg and some lengthy ground running at Cape Town. Johnnie Walker came with me for the trial lasting one month.

This commitment clashed with the start of training for the contractual course, which began on the Filton training simulator at the same time. The flight training took place on 206 (GBOAA) in May and was not a great success due to a high failure rate of 33 per cent. The course included Captains Walpole, Myers, Budd, Butt, Orlebar, Benwell, Dyer and Maynard. Our training may not have been as good as it should have but the real problem was some of the trainees themselves. The selection process failed to sift out one or two who were marginal on their past record. If a pilot has been stuck on one type for many years, some do have a problem when a major change is needed.

The Concorde conversion course remains an arduous grind with about six weeks of ground control before the simulator phase. The Concorde is not the most forgiving aircraft as everything happens much more quickly necessitating the pilot being ahead all the time, otherwise it turns round and bites. BAC were not able to fail any trainee without their performance being witnessed by Captain Norman Todd, BA Flight Training Manager, Concorde, who had the unpleasant task of telling unsuccessful candidates. When BA took over their own training, high failure rates continued until a more thorough selection process was introduced.

I went on a tour to the Far East in aircraft 203 with Jean Franchi and Gilbert Defer more as the senior man from BAC rather than part of the crew. The trip covered Bahrain, Singapore, Manila, Hong Kong, Jakarta and Seoul. Two flights were made from Manila to Hong Kong and back taking Madame Marcos on a shopping expedition. The approach into Hong Kong from the north was quite exciting as the approach is made at right angles to the runway until about 300 ft when a rapid right turn is made on to runway heading. Concorde made a hell of noise doing this but nobody seemed to mind. The flight time between Hong Kong and Manila was only one hour as there were no supersonic restrictions, as the route was over the sea all the time. The flight from Seoul to London was done in one day via Singapore and Bahrain in eleven hours forty minutes, and thirty-minute stops at Singapore and Bahrain. We were all very tired at the end of the flight and it convinced me that two long sectors, even in Concorde, were enough.

After this trip, I spent a couple of days in Gander with the same crew proving the modified landing gear on the Gander runway which was quite rough. The modification, which was eventually fitted to all Concordes involved a two-stage main oleo. This provided a vast improvement in aircraft response to rough or undulating runways and was accepted as providing the solution to the previous problem.

At this stage 201 and 202 were assigned to supporting in-service aircraft which started another argument as to which aircraft should be retained in the longer term. There was no doubt in my mind that engine and intake problems were the most likely areas to require attention, both of which were BAC responsibility. This view was not shared by anyone else at the time on the basis that systems of Aerospatiale responsibility like auto-flight control, hydraulics, wheels and brakes normally produced the most problems in service. Furthermore, the identified development items which extended furthest on time-scale were Aerospatiale areas – Category III Autoland, autopilot improvements, carbon brake improvements and simplifications.

All of this favoured the retention of 201 but, personally, I believe the future of Fairford played a major part in the decisions. To enable it to fly with either Mk602 or 610 engines like 202, 201 required a major update. The Mk 602 was used during development and the Mk610 became the production aircraft engine. The Government's decision to limit production to sixteen aircraft had serious implications for the Commercial Aircraft Division, which had to be addressed by the directors before the end of 1975. Manpower reductions were required at all of CAD's factories and the question of Fairford was directly linked to the decision. The hard fact was that Fairford was going to run out of work and the Flight Test Organization and Facility could not be sustained on the back of production flying beyond some period in 1976. There was a fear in the minds of some that arguments would be put forward to keep Fairford if 202 was retained as the only support aircraft. The situation was not helped by the fact that at the time of decision making the last four production Concordes had not been sold. Attempts to conclude deals with Philippine Airways and Iran Air were not producing very likely prospects.

The only way in which Fairford could have survived would have been the imminent appearance of a new project but even this would have left a gap. Like all Concorde funding, Fairford was paid for by the Government, who made it clear to BAC through the Concorde Management Board that they could not see any justification for Fairford beyond the middle of 1976. Detailed discussions between BAC and the Concorde officials took place over many months. I had to point out the difficulties of operating Concordes without the availability of Fairford due to the limitations of Filton, whereas Toulouse had the advantages of an integrated facility, excellent runways and above all another major project in the form of the Airbus A300 with more to follow. I saw the whole scenario as the demise of the British civil aircraft industry in the form that most of us had known it.

As announcements of the closure of Fairford were made, emotions rose very rapidly. Every so and so and his aunt became involved asking questions about the basic decision, ably fanned by some of the BAC staff, one of whom made a personal representation to the Earl of Kimberley. Dates changed slightly but the overall situation remained the same. I was instructed to take a much reduced flight test department back to Filton to complete the remaining production aircraft while retaining the ability to use Fairford for two to three weeks during the production testing phase, so that at least one maximum weight take-off could be carried out. It was also agreed that the ICL 9030 computer which was the property of the Department of Trade and Industry would also be retained for an unspecified period in its own special building.

Some people viewed the pending nationalization as a further complication but this really had no bearing on the future of Fairford as BAC management were expected to behave in a responsible manner with or without nationalization. Anyway, nationalization had already been indicated in the Queen's Speech in 1974 and the inevitable had to be faced. The immediate effect as far as I was concerned introduced the Chairman designate's interest to the Fairford scene.

The trade unions took the decision very badly and immediately blacked production aircraft 208 in spite of the fact that meetings held in November 1975 between representatives from each CAD establishment and the Managing Director, John Ferguson-Smith had pointed out that the lack of programme commitments for Concorde would not justify the continuation of operations from Fairford beyond the end of 1976. The closure meant about 204 redundancies. The unions raised a lot of questions, as one might expect, and produced a comprehensive report on the subject which made a case for retaining Fairford until 1980. Lord Beswick was warned by the Chairman of the Joint Shop Stewards' Committee that unless satisfactory answers were produced by 4 June, i.e., one week from the date of the closure announcement, then a work to rule on aircraft 208 would commence. The work to rule quickly changed to a complete blacking on instructions given by the BAC Combined Shop Stewards' Committee. The assistance of the Engineering Employers' West of England Associated was requested and used to resolve the dispute, which delayed delivery to BA until the end of September 1976 instead of June 1976. Aircraft 210 flew in August and was delivered in December.

Fairford was officially closed on 30 November 1976 and a much reduced flight test team returned to Filton. Some key staff were lost in the process and I recall being sent for by Frank Beswick to explain the background. I was somewhat surprised to find the Earl of Kimberley present as well. Lord Beswick's position as a Minister in the Department of Industry, under Wedgwood Benn as Secretary of State and Eric Heffer as Minister of State, was obviously created so that Lord Beswick could be named as head of the new aerospace company as and when it was formed. He was a most approachable man and I always found him very friendly. However, none of it altered the decision concerning Fairford.

The official announcement on the closure of Fairford was made by me at the end of May 1976 following numerous interchanges between Government and BAC including a debate in the House of Lords. Many of the concerns expressed were genuine and showed a strong desire to see the UK hold its own on the Concorde and any further developments. Unfortunately, much of the debate failed to recognize the fundamental point that CAD were running out of work and that Fairford could not exist without a continuous flow of activity and that it was not just a question of cost.

For all my understanding of the situation, I still felt pretty sick deep down at the prospect of breaking up a wonderful team who had laboured so hard for nearly seven years since first flight in 1969. If the BAC 3-11 project had survived the situation might have been different, but the flight development of such an aircraft at a separate base from one of the factories was bitterly opposed by a number of directors.

Roger White-Smith had already gone to Filton as General Manager, but had to retire for health reasons. Bob McKinlay returned to the Design Office as Concorde Design Director. Michael Crisp joined BAe Guided Weapons after over thirty years in Flight Test and Roy Holland of similar vintage went to the Military Aircraft Division. Mervyn Berry took over many of Roger White-Smith's responsibilities at Filton, while the Technical Staff were headed by Mike Bailey. On the air crew side, pilots John Cochrane, Roy Radford, Peter Baker and Flight Engineers Dennis Ackary, Peter Holding and Alan Heywood, and myself and Alan Smith, one of our chief 1-11 training pilots, took up residence at Filton. Here the arrangements were fundamentally different in that the aircrew and flight test technical staff were located in separate buildings, some half a mile apart. Eddie McNamara and Johnnie Walker retired, while Hurn remained more or less intact. I have laboured the closure of Fairford purposely because it identifies the parlous period that the civil side of BAC's interests were entering.

On the flight test side, BAC/BAe made a strong bid for the flight development of Airbus A320 a year or so later on, but it was difficult to produce a convincing case in favour of Filton compared to Toulouse, which had the A300 and A310 already and was expanding at a phenomenal rate. Needless to say, the programme took place at Toulouse.

EIGHTEEN

Nationalization and Back at Filton

The nationalization of BAC and Hawker Siddeley Aviation and Dynamics was a long and tedious process lasting several years, which made life very difficult for those running the companies, and created an unsettling and disruptive atmosphere for all members of the staff, for customers and international partners.

The first positive move against aerospace went back to 1973 when it was listed as a 'prime candidate' by Labour for nationalization and was approved by the National Executive in 1974, the same year as Labour was returned to power. As a result nationalization occupied the last two years of Sir George Edwards's chairmanship and the one year, four months of Allen Greenwood's term. I was obviously not involved directly in any of these protracted activities but I was well aware that it was a nightmare time for BAC and its parent companies Vickers and GEC.

Nationalization featured in the Queen's Speech in October 1974 and was clearly going to happen, but it took until the end of April 1977 for the Bill to be finalized and passed for royal assent, with Vesting Day on 29 April 1977. The Bill itself was considered to be a poor one in its original form created by Wedgwood Benn, and many amendments were made but the end result, although bitterly attacked, was the merger of BAC, Hawker Siddeley Aviation, Hawker Siddeley Dynamics and Scottish Aviation to form British Aerospace.

Lord Beswick was announced as Chairman of the Organizing Committee and Chairman designate of the British

Aerospace Board. The remainder of the Committee was not finalized for some time as several key members of the aircraft industry did not wish to be involved in a nationalized corporation. Sir George Edwards had already retired and so went one of the really great pioneers of British aviation, Sir Arnold Hall remained with Hawker Siddeley Group, Sir John Lidbury, Air Chief Marshal Sir Harry Broadhurst and Geoffrey Knight were all unwilling to continue, while Allen Greenwood, now Chairman of BAC, was very torn between going and staying, but after much persuasion he agreed to stay for a time as Deputy Chairman designate.

The original members of the Organizing Committee were announced as Mr L.W. Buck, Chairman of the Confederation of Shipbuilding and Engineering Unions; Mr G.R. (Sir George) Jefferson, Chairman/Managing Director BAC's Guided Weapons Division; Dr A.W. (Sir Austin) Pearce, Chairman of Esso; Mr Eric Rubython, General Manager HSA; Sir Frederick Page. Air Chief Marshal Sir Peter Fletcher and Mr B.E. Friend joined later. On Vesting Day, the Organizing Committee became the British Aerospace Board.

British Aerospace was immediately divided into two groups as had been anticipated. The Aircraft Group was headed by F.S. (Sir Frederick) Page and the Dynamics Group by G.R. Jefferson. On the aircraft side, Alec Atkin took on the role of Managing Director (Military) and Jim Thorne became Managing Director (Civil). John Ferguson-Smith became Deputy Managing Director (Civil) and Marketing Director Civil Aircraft Group. J.T. Stamper took the position of Technical Director. This led to Mick Wilde being promoted to Managing Director Weybridge/Bristol Division and I took over his responsibilities as Concorde Project Director. All the other appointments are too numerous to list but readers can be sure that there was no shortage of fancy titles.

The Aircraft Group housed itself at the old HSA HQ at Kingston and Guided Weapons set up at Stevenage. Every three months each aircraft divisional board, of which there were six, presented itself to Freddie Page and Eric Rubython, who ran the meeting for what might be politely termed 'a severe roasting' of their performance, while other members of the Aircraft Group Board looked on. I cannot say that these occasions impressed me very much as a method of management but they worried Mick Wilde tremendously. Mick was so conscientious that these events took a lot out of him and, when coupled with the eventual closure of Hurn, probably contributed to the eventual breakdown in his health, from which he sadly failed to recover.

The organization in both aircraft and dynamics groups produced 'a workable set up' and fulfilled Sir George Edwards's stated wish that this was his hope. In 1981 the scene altered yet again with a change of Government and BAe became a semi-private/semi-public company under Sir Austin Pearce. Nationalization was a long drawn-out affair which I do not think anyone on the industry side enjoyed with much relish. The final outcome coincided with the return of the remnants of Fairford to Filton. Each year continued to be full of what I regarded as very significant events and 1977, 1978 and 1979 were no exception.

The amount of our Concorde flying dropped off for the first quarter but picked up with some flight tests on a revised intake law installed on 202, which was also used for training the CAA Inspector assigned to BA, John Oliver and the Head of Accident Branch, Geoff Wilkinson. The twelfth production aircraft 212 (G-BOAE) came on the scene for production clearance and I was delighted that we went supersonic on the first flight.

The modified main landing gear was fitted to aircraft 201 and I went with Jean Franchi to Gander for a few flights as the Gander runway was sufficiently rough to use as a datum. We were only there for two days but we managed eight flights in quick succession. Once back at Filton, I had to re-validate my VC10/Super VC10 as the tanker programme for the RAF was beginning to move forward and the various aircraft had to be ferried in from Stanstead or Nairobi.

A number of minor performance improvement modifications were made to 202 by introducing extensions to the rudder and elevon trailing edges, changes to the rear ramp leading edges and further revision to the intake laws. This package involved cruise performance measurements, engine compatibility tests and intake load measurements and some of them required the use of Casablanca for two to three weeks. Sharjah had built a new runway and we at BAe were asked to christen it. This was accepted very readily as overseas trips in Concorde were not be turned down. We routed through Damascus and received a great welcome at Sharjah where the whole party were given watches and a few received magnificent swords as well. Once again Sir Geoffrey Tuttle headed the BAe party.

The final destination and resting place for aircraft 01 was finally agreed to be Duxford but the Ministry had some difficulty in the supply of any engines for 01. They came in the nick of time as the Duxford runway was about to be shortened by the M10 motorway construction. The Duxford runway was not over long anyway (6,000 ft) so I had to get 01 there before it became any shorter. The aircraft was finally prepared for flight which took place on a Saturday with the bulldozers due on the Monday. A tail parachute was fitted to 01 for braking, and to the two prototypes, so I had no trouble in stopping in about 4,000 ft. Regular re-painting has kept 01 in excellent condition, as it has to sit outside most of the time and is a very popular exhibit in the hands of the Duxford Aviation Society.

Rolls-Royce came up with a new worry about engine surge due to over-fuelling by the engine control, which had potential structural implications on intake attachments. Such a surge would occur on the high pressure side of the engine, as opposed to what was normally experienced. A special overfuelling box was fitted to 202, so that surges could be introduced on a very carefully controlled basis. Fortunately, the measured loads were much less than predicted and no operating limitations were imposed on the certified flight envelope. These surge tests were particularly unpleasant as they felt so hard. 202 continued to provide 'in service' support mainly on intake work, engine investigations and intake law changes on a regular basis.

In due course a modification, providing much improved performance by means of a thinned lower lip on the intakes was fitted. This involved a recertification of the intake control system due to the physical and law changes. The usual series of surge tests to prove the margins were conducted by deliberate surges and pushovers up to M=2.17. Casablanca had to be used on several occasions to cover the most critical cases. This was all highly successful and the modification was offered for sale to both airlines, who both accepted. At the end of the programme I thought to myself that engine surges had become a way of life. I do not think anyone got completely used to them even though it was possible to feel the surge coming, as inching the intakes ramps manually produced rough running prior to the actual surge. This was the last major task on 202, but the idea that 202 could retire in 1976 had been exploded several times over and its outstanding career lasted until December 1981.

Another conundrum appeared over the 'white-tails' that were under construction namely 214 and 216 in the UK and 213 and 215 in France. Various attempts were made to place all five unsold Concordes: 203, 213, 214, 215 and 216. Strenuous efforts to lease 203 to Singapore Airlines reached the point of firm proposals, but the manufacturers could not commit themselves to providing all the flight crews for more than one year on sheer grounds of availability. SIA wanted a four-year commitment. There were also problems regarding maintenance at Heathrow and the provision of spares. The flight crews would be BAC and Aerospatiale personnel. A counter proposal was also made by BA who succeeded in reaching an agreement with SIA for a joint service London–Bahrain– Singapore using a BA aircraft, one of which had SIA colours painted on one side of the fuselage. The BA/SIA operation ceased in 1980.

Any idea that manufacturers' crews might be used on a BA aircraft was soon exploded by Captain Norman Todd, who stated that any of us would need at least six months training to come up to BA standards. I knew what he meant as our crew operation was not as rigid as BA, but it could have been put more tactfully.

At one stage I suggested to the RAF that they might use two aircraft as supersonic targets. I suggested that I would just appear but this was met with 'Do not do that for Christ's sake', but the idea was met with some interest and I actually flew 202 so that the RAF could reach an assessment. I was attacked by fourteen Lightnings head on, who were heading north over the North Sea. I did not see any of them except the one that 'missed'. All the others 'got me' with a missile fired from several thousand feet below, which was about the end of this little venture. A number of other airlines were approached but never reached any agreements. In the end the aircraft were sold to BA and Air France for a 'nominal figure'. This was so low that I once upset the Concorde Assistant Secretary DTI, Bruce McTavish, when I told him that I was thinking of buying one myself.

An interchange agreement was made with Braniff whereby BA and Air France services to Dulles, Washington continued on to Dallas subsonic flown by Braniff crews. A few minor modifications had to be applied to meet FAA requirements, which resulted in the award of a United States type certificate in January 1979. The aircraft carried US registration numbers for this purpose and the Braniff crews were trained between BAe at Filton and Aerospatiale at Toulouse. I think that Braniff hoped that further routes would become available in due time, especially to South America. Like many other good ideas it floundered.

Braniff crews found the ground school learning techniques at Aeroformation (Aerospatiale flight training subsidiary at this time) much more in keeping with their own methods than the more stereotyped although excellent ground training at Filton, whose pattern was closely linked to passing the CAA examinations through the multi-choice answer system. I hate this system as it requires a great deal of mental retention of facts and figures which are unnecessary in my opinion but it means that a clerk can correct the answer papers. The Braniff group were a delight to work with and only one captain found the arduous course too much and returned to B-747s.

A concerted effort to interest Federal Express led to a complete feasibility study. This identified numerous changes to the passenger cabin which was divided into freight 'zones' and alterations to the air conditioning. A supersonic parcel service had many attractions but producing a meaningful utilization within night curfews and other constraints was not so easy.

At least serious attempts to sell the unsold aircraft were made at all levels, albeit without success. BA finished with seven and Air France also followed to a fleet of seven. Over the years BA could have done with another aircraft while Air France have operated on the basis of six as one of the Air France aircraft was damaged some years ago in a heavy landing and was not put back into service and has subsequently been scrapped.

NINETEEN

Flying a Desk

In 1980 Jack Jefferies retired and I took over as Director and General Manager at Filton. There was no formal handover, which I thought there might have been, so I had to pick up the reins as best I could. The task also included taking over a number of projects on the Filton site at that time – namely VC10 tanker, F1-11 major overhaul programme, BAe146 centre fuselage construction and several others. Of these the 146 fuselage programme seemed to have priority thanks to the presence of Aircraft Group and the VC10 had gone to the bottom of the pile. I spent several years getting the tanker programme back on its feet, but at a cost.

The F1-11 work had started as a simple replacement of the pyrotechnics associated with the crew escape capsule and was not very labour intensive. I remember being attacked by Admiral Sir Raymond Lygo, when he became Chief Executive of BAe, about the hangar space being used for such a small return. I stood my ground and pointed out that this was merely the start and indeed the F1-11 blossomed into one of the best money-making projects in the group. All the F-11s in the UK were subjected to a complete overall/rebuild at Filton instead of being flown back to Sacramento, USA. Some special facilities were constructed at Filton and paid for by USAF.

The matter of Concorde in service support costs arose once more under the scrutiny of the Industry and Trade Select Committee and some weeks were spent preparing statements, costs estimates, for example, and took place at the end of 1980. BAC's final statement was submitted to the Clerk of the Committee following several meetings with Concorde officials and our own legal department. BAe were instructed by John McEnery and Bruce McTavish of the DOI Concorde office not to include financial data as they would tie all cost information into one document.

On 29 January 1981, BAe represented by Mick Wilde, John Dickens and myself went in front of the Committee under Sir John Kaberry. Other members were Messrs Carlisle, Cockerum, Crowther, Emery, Foster, Kerr, Maxwell-Hyslop and Mikardo. BA were represented by Stephen Wheatcroft, Captain Brian Walpole and Peter Brass and Rolls-Royce by A. Newton, P. Torkington and A. Warrington.

When it came to our turn, we received a warm welcome from the Chairman who complimented Mick Wilde on a beautifully written and presented statement, but immediately turned on Mick rather like an enraged bull and said 'Is it not a remarkable thing that in a memo-randum submitted as a contribution to an enquiry into the operating and continuing costs of Concorde nowhere in that memorandum is there any mention of any sums of money? The £ sign is nowhere to be seen.' Mick responded that the DOI had considered it best to submit all costs from every source as an aggregate in their submission. A great many searching questions followed during the session and it was interesting listening to the other parties as well. Other sessions including one involving ministers also took place, I believe. The outcome as I understood it was that there was no apparent advantage in cancelling the Concorde support against continuing.

However, the matter did not rest there for very long before HMG reached a decision not to continue paying for all Concorde in-service support costs. This raised a number of major issues and produced a scheme whereby all Concorde assets and costs should be taken over by BA for a quoted sum of money. The French took some convincing about BA becoming the UK banker, operator and everything else. Under this arrangement all costs incurred by BAe and Rolls-Royce would be met by BA and, much to the British officials' surprise and amazement, BA accepted the offer, picked up the ball and have run with it ever since.

Being the General Manager was a new experience for me, requiring me to spend a lot of my time on industrial matters with the unions, particularly addressing the problem of differential pay between sites. This issue was complicated by the fact that, in order to survive, Filton had acquired manufacturing work from several other BAe factories like Warton where the Tornado was produced. At Warton hourly pay rates were much higher and this eventually led to a major problem.

There was still the occasional opportunity to fly with Roy Radford who had succeeded me as Chief Test Pilot and we flew the last flight of 202 together on Christmas Eve 1981. An occasional trip with BA was a very welcome break in a life which had become very different to what I had experienced for most of my long time within the company. The trips with BA were not only fun but very rewarding in terms of witnessing Concorde being expertly used in the role for which it had been built.

I had a very close relationship with Brian Walpole, who had become General Manager of the BA Concorde Division and who made a huge contribution to the success of Concorde in BA. The earlier rather unwelcome approach to the aircraft was transformed when Lord King and Sir Colin Marshall took up the reins. Their support of Concorde coupled with Brian Walpole's energy, enthusiasm and foresightedness created an operation that has become very successful. Jock Lowe helped Brian Walpole a great deal and later became Director of Flight Operations for BA before becoming Commercial Manager, Concorde, a position which he holds today while remaining the longest-serving Concorde pilot.

I also used to fly with BA at the SBAC Show as my presence on board eliminated any requirement for a rehearsal, which would have been difficult to fit in and was expensive. I flew with Brian Walpole, Dave Leney and John Cook on separate occasions and usually got a chance to re-acquaint myself with Concorde on the way back to Heathrow.

All this helped me to acclimatize to 'flying a desk'. It had been a gradual process for ending my active flying career and I got used to it quite quickly. I was also helped by the fact that there was not much happening on the flying side in the Commercial Aircraft Division for the time being at least. I have to admit that I have always missed flying even in advancing years, but really Concorde spoilt me from wanting to fly any other aircraft.

I look back on my time as Director and General Manager at Filton as covering a period of survival for the Weybridge-Filton Division of CAD. Work had to be grabbed with every opportunity but the threat of redundancies kept rearing its head. The indirect manning situation at Filton always looked worse than it was because the aircraft side provided all the services, like police, fire brigade, telephone operators to Dynamics who were also located on the Filton site and who paid for the services provided. Dynamics were much more successful at the time than the aircraft division and I was often taken to task by Sir Raymond Lygo for having too many 'indirects' and he was not very interested in my explanations. Occasionally I would get a message that he was in one of the Department's offices, talking to staff unannounced. By which time he had probably left but this was followed by a hastily dictated note asking for answers to numerous points that had been raised. I always said to myself, 'I bet that he never got away with this when he was in the Navy'. Mass addresses to all personnel on the site also used to take place at periodic intervals.

Mick Wilde worked incredibly hard to keep Weybridge-Filton viable but in the end he was left with no alternative but to close the Hurn factory, which operated as part of Weybridge. This took a huge toll on him as laying off several hundred people was not a pleasant task for anyone and soon afterwards he was taken ill at a Board Meeting at Kingston and rushed off to St George's Hospital in south-east London. I went to see him there and was very unimpressed as well as concerned by his extremely frail condition. He was transferred to Frenchay Hospital near Bristol about a week later and I saw him again but not for long, as he unfortunately died. We were all incredibly sad as Mick was such a likeable person, apart from being a brilliant engineer. He was probably too nice for the job he ultimately had to do, as being so conscientious and very sensitive took a heavy toll on Mick, both physically and mentally.

After Mick died, John Glasscock took over at Weybridge and Filton as he was already Chairman of the Commercial Aircraft Division. This move created an opportunity to introduce someone young and up-and-coming to be groomed for a senior management position leading to higher office. Bob McKinlay, who had gone from Concorde Design Director to take over the BAe 146 at Hatfield, was chosen. I had less than two years left before the mandatory retirement age of sixty-two, so I was perfectly happy to work with Bob and the fact that he had worked for me at Fairford did not bother me at all, although some thought that it might. This was just the start for him in his rise to become Chairman of the BAe Airbus programme.

Unfortunately, we soon ran into serious trouble. I had written to John Glasscock advising him that the pay differential between Filton and Warton was of such proportions that I anticipated major industrial action. I recommended what I thought was needed and it did not come to an enormous amount of money, but the knock-on effect throughout the Corporation was considered by BAe HQ and the award was refused. I went into Filton early one Monday morning following a tip-off that industrial action was imminent. I had sat in my office for a short time when the thought occurred to me that I was not best placed in case I got trapped in the building, so I went out to my car. At this moment, a lorry full of benches and tables drew up with one of the stewards who I knew very well. I greeted him and he responded 'We are taking over'. So, without further ado, I was off in a cloud of smoke towards the airfield. As I reached the bottom of the hill leading down to the airfield, I saw a small space open in a barricade across the road, though which I could still drive my car. I went straight to the Flight Operations buildings in time to see the whole Filton factory taken over. Fortunately Flight Operations had one or two direct telephone lines, which were used to set up emergency headquarters. From here Bob McKinlay, Mervyn Berry and I communicated with the unions, full-time officials, Employers' Federation and the police.

We remained in this situation for several weeks, during which a very heavy storm hit Filton one night, which caused me to be particularly concerned about possible damage to the large hangar. After due deliberation with the unions, I was allowed to go through the barricade and taken under escort to inspect the hangar, which turned out to be all right. As I went through the barricade, the guard lifted his cap and said 'Good morning Mr Trubshaw'; I felt somewhat peeved at having to grovel but I had little alternative as an inspection of the hangar was essential.

All disputes have to be settled eventually and this was no exception. We discussed various suggestions of how to take the site back including a commando-type attack by the police and all sorts of other ideas, but fortunately a suitable settlement with the unions was reached. It was a sad affair and it cost BAe a lot of money, programmes were delayed and the publicity did not help anybody. My small reward, which was not much more than a few packets of fags in value, was dwarfed by the overall result, which included repairing some plant that was damaged during the dispute.

We got back into our stride as quickly as possible and prospects began to look a lot brighter. More Airbus work, VC10 tanker flying forging on and a large expansion in the F1-11 programme, which included the provision of the special facilities to support the programme that I have already mentioned. So, when the day came for me to retire in January 1986, I left in good spirits with a splendid Atco motor mower as a parting gift.

I often look back on my years with Vickers, BAC and BAe with a kind of gratitude. Not for the money they paid me over thirty-six long years, although I enjoyed my work very much, but for the many deep and long-standing personal relationships I made throughout the course of my working life.

TWENTY

Life After British Aerospace

I was determined to continue working and I had had some discussions to that effect in the last few months at Filton. Two of these brought immediate results, one was collecting Concorde items and memorabilia for Wensley Haydon-Baillie who had a private museum collection at Southampton and the other was with SAC, a company providing design services to main corporations like BAe and Rolls-Royce.

Wensley Haydon-Baillie and his older brother Ormond had started a collection of vintage aircraft, Spitfires, Mustang, Blenheim, Lockheed T-33 and also an interest in naval activities, especially the Brave Class Boats built by Vosper in the late 1950s. I had met Ormond some years before on a visit to Edwards Air Force Base but he was tragically killed in their Mustang at an air show in Germany soon afterwards.

After Ormond's death, Wensley sold most of the aircraft collection and concentrated on restoring the one Brave Class Boat built for civil use and subsequently owned by Mr Niarchos. Wensley also held a passionate love for the Concorde and tried to buy 202 from British Airways. His proposal was drawn up in a firm and perfect contract, whereby he would house 202 in a specially built hangar on the edge of Filton airfield and would permit BA to take spares off it as and when required. This nearly succeeded but floundered at the last minute on a legal issue once the BA lawyers became involved. The lawyer could not fault the contract but pointed out that if Wensley reneged in any way it would be necessary to take the matter to court which would take time. This was unacceptable to Brian Walpole.

Undaunted by this set back, Wensley set about acquiring a fascinating Concorde collection with my help. I found bits and pieces in the strangest of places. BAe provided the main source, as they desired to get rid of old and supposedly useless items which had been part of Concorde development. From these acquisitions, it became possible to make some swaps. Some of the actual nuts and bolts that I acquired for him were subsequently considered usable. I obtained three Olympus 593 development engines and other engine parts in exchange. We also bought several Avon engines from the MOD and then swapped ten of them for ten TSR2 engines, which I found under some brambles and scrub at MOD Shoeburyness. This was an extraordinary experience as I got a tip-off from a friend of mine at Rolls-Royce who told me that these engines were at Shoeburyness waiting to have rockets fired at them in order to establish patterns of engine damage. I contacted the RAF Commander, a Flight Lieutenant Angela Kirby, who was the Weapons Officer, and she took me out to the 'engine store'. This turned out to be an area of brambles, scrub and bushes among which were lying various engines including eight TSR2 engines (two more were found later) one of which had never been out of its box. I soon established that firing at an Avon engine would be just as good, so I arranged a direct swap of ten TSR2 engines for ten Avons, which we sent to Pendine. Also, from MOD surplus, we acquired a Vulcan engine, thus providing the Haydon-Baillie collection with the whole range of the Olympus, starting with the Vulcan then developed for TSR2, then re-vamped for Concorde. Another tip-off revealed that all the original prototype and pre-production drawings were about to be destroyed, so I got my hands on them, about 77,000 to be precise. Gradually, the collection built up a fascinating package of Concorde items of all sorts including wind tunnel models and other test specimens. I acquired a working model of the intake and one of the original airframe models. Other shapes came from RAE Bedford.

Wensley Haydon-Baillie also had a collection of Rolls-Royce cars consisting mainly of the Phantom III and Flying Spur. This enabled his collection to show Rolls-Royce on land, on the sea and in the air, which no one else could do. *Brave Challenger* as restored was the most fabulous example of British workmanship that I have ever seen. It was quite beautiful and it did over 60 knots to remind any doubter about its unique origin as the only civil boat built to Brave Class specification. More recently most of the collection has been moved to Wensley Haydon-Baillie's ancestral home, Wentworth Woodhouse in Yorkshire.

I had had a long association with SAC, a company which had started some twenty-five years earlier by three ex-Rolls-Royce apprentices – Smedley, Allard and Creer. I had a fair amount of contact with Peter Allard while I was still with BAe. When I retired he kindly offered me a consultancy with his firm to help acquire additional contracts from the aircraft industry. SAC subsequently merged with Ricardo, who have recently sold off the aviation design side to Inbis.

My next move was to accept two approaches from Joe Lewis, owner of A.J. Walter (Aviation) Ltd. and the other from C.T. Bowring & Co., the insurance group. I became a director of A.J. Walter (Aviation) Ltd. and had a long and happy association with Angus Whiteside, Christopher Whiteside and Toby Silverton, who set up a parallel organization in Orlando. This is now a separate company called Jet Spares Inc. Similarly, I thoroughly enjoyed the ten years as aviation consultant to Bowring Aviation Ltd., executing introductions to companies who might be persuaded to insure with Bowring. This proved to be a difficult task, as I soon learnt that big companies tend to stick to the same insurance brokers for years, some of which are founded on the back of personal relationships.

I then made an approach to the Air Worthiness Authority

in order to determine if there were any vacancies on the board. This turned out rather better than I hoped, because I got an immediate invitation to join the Civil Aviation Authority as a part-time member, firstly under Christopher Tugendhat (now Lord Tugendhat) and secondly under Christopher Chataway (Sir Christopher). I had to get through one or two interviews first with the Department of Transport Ministers and officials, which were all successful. I was made very aware of the hazards of what is now called 'sleaze' and was tactfully pressed to sell my British Aerospace shares, all 350 of them I recall, just in case of any conflict of interest that might arise.

Being on the CAA Board for six and a half years was rather like looking at the aircraft business from the other side. I was slightly surprised that of the other board members only one, Rex Smith who was an old friend, had any immediate knowledge of aviation, but I think we had a good team and Christopher Tugendhat was the most excellent chairman. He brought in Tom Murphy from BP as Managing Director and this proved a huge benefit thanks to Tom Murphy's high-calibre skill. One of the duties of part-time Board members was to hear appeals made against the CAA under what is known as Regulation 6 of the Air Navigation Order. These appeals usually involved individuals who had had their licence revoked for one reason or another. At the start, I thought some of the CAA licensing decisions were high handed to say the least and ruled in favour of the appellant. This may seem a strange arrangement where a Board member was ruling against his own organization. However, it has to be remembered that the part-time members are not appointed by the CAA but by the Government through the Department of Transport. Rex Smith and I had many discussions with the licensing people and I believe that an improvement took place in the way that licensing problems were tackled.

I was also a member of the Capital Expenditure Committee, whose main activity was dealing with new items for the National Air Traffic System (NATS). It was quite clear from the way the supporting papers were drawn up, that NATS had previously been facing a hostile audience and there had been too much opposition to re-equipment. Tom Murphy changed all that and a great deal of progress was made in the time I was at the CAA. Many of the projects were extensive and expensive and managing them to time and cost was often a nightmare and from what I still hear and read the situation has not changed that much, although these days software is always blamed for almost anything.

After retiring from the CAA, I was offered an appointment by Brian Calvert, who was by then Managing Director of Hunting Aviation Services. Hunting had become involved in the calibration of navigational aids and had developed a much more modern and simple system than used by the CAA. I had had some exposure to the system while at the CAA when Hunting applied and were granted CAA approval for their system by the Safety Regulation Group. A few eyebrows were raised at another organization carrying out calibrations, which had always been done by the CAA Calibration Unit and a question was raised regarding safety, but this was easily disproved. My association with Hunting, which went on long after Brian Calvert left, ended in 1997 when my particular areas of interest were put up for possible sale.

Another opportunity of ever-growing interest arose when I was asked by Captain Jock Lowe, Director of Flight Operations BA, to join the Flight Operations Safety Board to which I readily agreed and am still a member in 1998 under his successor Captain Mike Jeffery. I have found this to be one of the most interesting activities of my aviation life. The purpose of the Board is firstly to review all incidents of note and to establish that the action necessary to prevent a reccurrence has been taken. BA have a very sophisticated flight recording system, which highlights any unsatisfactory trends or provides the ability to examine in great detail any individual occurrence, all of which is to help improve safety standards. It is not only the technical content that is impressive but I find the sheer professionalism within BA to be the most reassuring feature that any prospective passenger could have. I wish more people knew what goes on behind the scenes.

Very recently, British Mediterranean Airways, who operate under a BA franchise and operate Airbus A320 to the Middle East in the main, have invited me to be a member of their Flight Operations Safety Board. An invitation which I have accepted with great delight.

Then there is always the Society of British Aerospace Companies (SBAC). I have been a member of their Flight Operations Committee for forty-seven years, which must be a record. The main task of the Committee is to deal with the bi-annual flying display at Farnborough in setting the flying regulations and determining the actual display in content, timing and order of events. The Committee also addresses all matters concerning flight operations that may affect the aircraft industry and is becoming more and more involved with European Community aspects.

Large air shows are not as easy to manage as the eventual spectacle may imply. There is an ever growing need to raise safety standards from every point of view. Over the years there have been dramatic changes to the flying regulations incorporating lessons learned from actual events, some of which have resulted in tragedy. At Farnborough, the flying is overseen by a Flying Control Committee, comprising members of the Flight Operations Committee, SBAC under the Chairmanship of an RAF Group Captain, who answers to the Ministry of Defence. In 1998 Farnborough is still a Ministry of Defence Airfield and the Chairman is a retired Group Captain. When I first flew at Farnborough in 1950, I do not remember there being any rules except that any participating aircraft had to have flown for ten hours or more. In the early 1950s there was a tragic accident when

John Derry's de Havilland 110 broke up, depositing large pieces of wreckage, including the two engines, into the crowd, a large number of whom were killed or injured. I was standing with John's wife Eve and John Cunningham at the time, but we continued the flying display after a short delay and Eve stayed until the end, an act of tremendous courage.

After this, flying towards the crowd was not permitted beyond a defined display line. Height limitations have been introduced and more particular emphasis has been placed on recovery from manoeuvres in the looping plane. In this case there have been far too many aircraft 'tent pegging' into the ground because recovery from a looping manoeuvre has been initiated at too low a level. Speed limitations also apply, whereas when the Swift and the Hunter first appeared the highlight of the show was these aircraft planting a 'sonic boom' in the middle of Farnborough, as they broke through the sound barrier.

In spite of strenuous efforts and months of preparation sending out the show regulations and so on, accidents do happen. At Farnborough each participant has to rehearse their display for good and bad weather in front of the Flying Control Committee satisfactorily before he or she can display on the actual show days. The Committee has the right to refuse any display, terminate it at any point or withdraw the aircraft/pilot altogether if the Committee is sufficiently concerned. The latter does not normally occur but it did happen recently when the Committee considered a particular manoeuvre to be highly dangerous. Each time there is any mishap at an air show, there is a potential knock-on to all other shows and that is why it relies on everyone concerned to do their utmost to avoid accidents or even incidents.

Few spectators realize the organization that is put into a show like Farnborough, or the Royal International Air Tattoo (RIAT), Fairford. Once one show is over, it is time to start organizing the next. So much so, that the SBAC have a small permanent staff at Farnborough and a full-time

Director in charge of Exhibitions, Peter Taylor. The RIAT have a full-time Director, Paul Bowen and Tim Prince, Director of Flight Operations heading the sizeable permanent staff, who are supported by a further 2,500 or so volunteers for the actual tattoo.

I have left reference to the SBAC to end quite purposefully because it represents the coming together of the industry that I have cherished for many years. There is a board in the Council Room listing the names of all the presidents that speaks for itself as these are the great names of British aviation.

TWENTY-ONE

The Future

Will there be another supersonic transport? I believe the answer is 'Yes', but the timing of a new supersonic aircraft is not so certain.

Concorde has taught many lessons, all of which are available to the designers of the next generation in both the technical sense and the operational experience of over twenty years. There is additional background as well because Concorde itself would have been further developed if more than sixteen production aircraft had been built. Aircraft 17 was defined as the 'B' model and incorporated several potent changes. The wing was modified to include full span droop leading edges and extended wing tips. The power plant was improved by increasing the mass air flow by 25 per cent for take-off and about 35 per cent on approach. This would have been achieved by increasing the diameter of the L.P. Compressor replacing the single stage L.P. Turbine by a two-stage turbine and adding a bleed device to increase the surge margin of the L.P. Compressor. Acoustic treatment of the intake and exhaust systems was also proposed.

The aerodynamic changes benefited the lift/drag ratio, giving increased range and improved noise performance for the fly over and approach cases. The increased mass flow in the engine reduced jet efflux speeds. Thus the jet noise was substantially reduced as a result of the lower jet velocities. The hot part of the engine was retained and the turbine entry temperatures were not changed. The reheat could be deleted, an increase in thrust over the whole flight envelope, particularly in supersonic cruise and an improvement in specific fuel consumption over the whole envelope, especially transonic, were achieved. The modification package would enable a reduction in airport noise levels, an increase in range and a reduction in block fuel.

I believe the 'B' model would have resulted in a totally different situation regarding sales to airlines. But it is no more than history now because it never happened. Failure to develop has been the bug-bear of British civil aviation for all the years that I can remember. Serious studies have been undertaken over the last few years in order to determine the commercial viability of a second generation supersonic transport. These studies have been made on a European basis and indeed on an international level involving manufacturers of airframes and engines, research agencies and airlines.

Technology has made significant strides since the days of Concorde design in almost every field. New materials and avionics are examples, both of which lead to the ability to produce a much lighter transport relative to Concorde. It has been possible to lay down basic concepts of an aircraft providing about 300 seats and with a range in the order of 5,500 miles. The latter permits full access to routes in the Pacific. Speed requirements range between M=2.0/2.2 to M=2.6.

In order to have a datum from which to work, BAe took Concorde and applied the latest technology to it. This produced some interesting results by indicating 800–1,000 nautical miles increase in range or the ability to meet Stage 3 noise levels due to operating at lower weights. Boeing in particular have had a sizeable number of design engineers working on a supersonic project, while in the UK and France financial restraints have controlled the level of activity.

There is no doubt that the technology will exist to build an aircraft on the lines suggested but there are many other issues to be addressed, which tend to push the project out in time. There are two basic differences between US and European thinking. Europe favour an initial range of 5,500 nautical miles compared to the US who believe in 5,000 nautical miles initially. The airlines themselves support longer ranges of 6,000 nautical miles. Speed is the other difference, the US believing M=2.4 is required on the basis that the slight time advantage over M=2.05, which is the European position, can help to fit schedules more readily into airport time curfews and therefore assist aircraft utilization. Europe do not consider the extra speed as being a tangible benefit and, with Concorde experience behind them, are very conscious of the effects that a higher cruise Mach number brings in relation to higher operating temperatures and intake configuration. The latter is very important to anyone who wrestled with making the Concorde Intake System work satisfactorily. Choice of M=2.05 would permit the use of an external compression system like Concorde, where the shock system which slows the air to M=1.0 at the throat is all external and inherently stable. A mixed compression system could be used, but as part of the shock system is within the intake duct it is inherently unstable. However, it has the advantage of lower external drag providing the control system is fast reacting to stabilize the shocks. Instability can result in what is known as an 'unstart', which is violent and difficult to recover. The Lockheed SR71 Blackbird has this problem. At the higher cruise speed of M=2.4 the external compression system cannot be sensibly envisaged due to high external drag thus limiting the choice to a mixed system. The operating temperatures resulting from this speed have major structural implications on the materials used, necessitating more expensive solutions and on other items like seals, sealants, oil, grease and cable insulation.

Aerodynamic targets are much higher than Concorde requiring some 35 per cent improvement in subsonic lift/drag ratio in order to achieve 50 per cent better in supersonic L/D, the same specific air range subsonic and supersonic, structure weight 40 per cent better than Concorde and chapter 3 noise levels with some margins. A comparison to Concorde shows an aircraft that is some 50 per cent bigger, twice as heavy and with three times the passenger capacity. Dimensionally the aircraft will be considerably longer than the large subsonic varieties, which have yet to come but with a smaller wing span. The power plant will need to show a 12 per cent improvement in supersonic specific fuel consumption and 25 per cent subsonic over the Olympus 593, reduced jet velocity and a stringent emission index. Low nacelle drag is also required in the order of 7–9 per cent of total aircraft drag.

All the targets quoted are likely to be achievable in the next few years but I am convinced that the other challenges involved are going to be the most difficult issues.

Fundamentally the aircraft has got to be built and sold at a price that the airlines can afford to pay, thus permitting the sale of tickets at a price, albeit with a small surcharge applied, that the ordinary individual can afford. The effect of such an aircraft in a mixed fleet has to be considered and environmental issues will play an important role. Research into some of the environmental aspects has now reached a considerable level. A lot of attention has been paid to emissions, especially of Nitrogen Oxides (nitric oxide and nitrogen dioxide) known as 'Nox' into the upper atmosphere and their impact on the ozone layer. There is a strong suggestion that ozone depletion due to 'Nox' is much less than previously assumed. Indeed a slight increase in ozone concentration may occur but, if this is proved to be the case, convincing the world community will in itself be difficult.

One particular proposed difference to Concorde will be the elimination of the droop nose and visor. This arrangement is necessitated on Concorde by the high-pitch altitude on take-off and approach. The presence of high lift devices on the advanced aircraft will reduce the high pitch altitude but it is still likely that the pilots' vision is impaired. The solution proposed is to utilize artificial vision based on current military technology. Deletion of the weight and complication of a 'droop snoot' is an attractive proposition but there is no great enthusiasm from the airline pilots that have been approached; I share their views most emphatically. The idea takes me back to the metal visor on Concorde. God gave man two eyes and it should be possible to use them properly even on a supersonic transport. The matter of military technology reminds me of a story told to me by an American Air Force officer. We were discussing the Lockheed F-117 Stealth Fighter and the Gulf War, when he observed that the bombing system on the F-117 was so good and accurate that it was '. . . the only aircraft that we could allow downtown Baghdad'. Seemingly this system can see the target to an uncanny degree, therefore the extreme accuracy of F-117 bombing system would avoid hitting the civilian property.

Air transport has a history of progress and I am sure that the trend will continue. There are probably larger subsonic transports to come, bigger than the Boeing 747-400 and these may absorb the funds available for a time, but I am a great believer that a second-generation supersonic transport will eventually come after the year 2010 and led by the United States.

I would love to be involved again, but I shall probably not even be around when it happens. I often look back on my long life as a test pilot and I always conclude that, given the chance, I would do it all over again. I think that something gets into one's blood to promote this feeling. Now I see that Senator John Glenn, former test pilot and the first American in space, is about to embark on another space mission at the age of seventy-seven. I envy him and wish him luck. In our line of activity you need a bit of luck and I have had my fair share during thirty-five years of test flying.

TWENTY-TWO

Epilogue

A personal tribute by Sally Edmondson

Brian Trubshaw died at home on Sunday 25 March 2001. It was the end of an era, not just for our family but the world of aviation as a whole.

Brian was a truly remarkable man, intelligent, special, revered by his peers and loved by his friends. Quite how much people loved him became evident in the days following his death, when my mother and I struggled to keep up with the volume of telephone calls that flooded in from all over the world, and fully to appreciate their content. The calls came from truly devastated individuals, from top-notch, tough industry professionals to those who had known him in the village shop. They were followed over the next few weeks by hundreds of letters. We tried to respond to them all, with a little help from British Aerospace, but I think we failed miserably.

I was, without question, very lucky to have Brian as stepfather, but I think the world of aviation was lucky to have such a passionate driving force within the industry.

At the time of Brian's death, Mike Bannister was British Airway's Chief Concorde pilot. He aptly summed Brian up when he said, 'Without doubt his energy and commitment were vital to making the Concorde programme the success it was.' In the words of another British Airways spokesperson: 'Everyone in British Airways who worked with him could not help but be inspired by his continued enthusiasm and joy for Concorde.'

Buckingham Palace also paid tribute to Brian, describing him as close to the Royal Family and a particular friend of Prince Philip, who unhesitatingly wrote the foreword to this book. 'The Duke is very sad,' they said.

Brian was awarded the CBE in 1970 and was known as 'my Brian' by the Queen, who knew him from his days with the King's Flight after the war. Brian's flying skills had been rated as 'exceptional'. He joined the King's Flight in 1946, piloting George VI and other members of the Royal Family. He had so many fond memories of those days, especially when he was occasionally roped in for after-dinner games with the young princesses at Balmoral. He was very proud to have had the opportunity of serving the Royal Family.

I take all the credit for bullying Brian into writing his autobiography twenty-six years after he became my stepfather. Over the years I had heard him tell all these wonderful stories that I felt should be documented, but I have never considered that the autobiography did him the true justice he deserved. In my opinion, it failed to convey the *essence* of the man who devoted his working life to the furtherance of aviation and supersonic flight. The blame for this could be attributed to his co-writer not doing her job properly. In my defence, it was partly due to the fact that he breezed through the early chapters, as he could not wait to start on the seven chapters that highlighted his work with and his passion for his greatest achievement: Concorde. The other reason was that he was a staunchly modest and private man. When I first started pushing him to write his autobiography, he was completely incredulous, fobbing me off by saying, 'Who do you think would want to read it?!' But I kept pestering him and I am glad.

Brian was someone who met and kept friends. At 13 Brian and the late Richard Grant-Rennick were new boys together at Winchester. Sixty years on they were living in the same village, enjoying each other's company and the closest of friends.

Douglas and Joan Bader stayed with my mother and Brian many times and always for the Badminton Horse Trials. Douglas obviously admired Brian's achievements in the same way that Brian admired the man who had become a hero and a legend.

The late Vulcan test pilot Roly Falk knew Brian during their time together at Vickers. Brian was godfather to Roly's youngest son, Geoff. The Falk family were introduced into my life in 1977, when they were living in Jersey. Roly's wife Leysa rang Brian, wanting to arrange a hockey fixture in Gloucestershire. Brian said, 'I don't know anything about hockey! I suggest you speak to my stepdaughter Sally. She's sporty!' And they have been great friends of mine ever since.

There was no side to Brian. What you saw was what you got. He was a countryman at heart. He often said that, if he had not been a pilot, he would have liked to have been a farmer. During the years that followed his marriage to my mother, he was working very long hours on the Concorde project. When he was at home he unwound in the garden – not that he ever gave the impression that he was stressed by his job, because he was so passionate about it. It was much more stressful for my mother. The fear that all test pilots' partners know is just part and parcel of the everyday working life of the people they love. My mother and Brian spent quality time together in the garden, really enjoying themselves, my mother feeding and watering plants, Brian digging ditches and pulling out hedges and tree stumps until the light faded. When Brian wasn't flying, his second passion was gardening.

Brian was also a great golfer; he was once given a set of clubs by Bobby Locke, which we still have. Unfortunately his punishing work schedules left him no time to practise.

Brian's other great loves were his Springer Spaniels Charlie and Clay and his Shetland ponies Caspar and Senator. He cried when Charlie died, so I am not sure that he would have been cut out to farm livestock.

Brian was a man who never gave up on what he believed in and would give 110 per cent in order to convince his counterparts that he was right

His coolness in saving Britain's prototype VC-10 from disaster on an early test flight won him the Derry and Richards Memorial Medal for 'outstanding test flying contributing to the advance of aviation' in 1965. Structural failure had been threatened when an elevator section broke loose and the aircraft shook 'as though the tail was shaking the dog'. Brian could not read the instruments because of the violent motion, but broadcast to base the nature of the trouble in case he could not get back. He then managed to land the aircraft with only half the elevator control. He later modestly described this manoeuvre as 'one of my trickier moments'.

Three years earlier, Brian had been awarded the same medal for his work in the early 1950s on the Valiant jet bomber, on which he tested the delivery system for Britain's first atom bomb, the 10,000lb Blue Danube.

In 1962 the British and French governments signed an agreement to develop a supersonic transport aircraft, eventually called Concorde (the French having insisted on the final 'e'). The chairman of BAC, George Edwards, selected Brian as test pilot.

Development of the aircraft proved problematic, as costs rose from £140 million to more than £280 million. But, despite some political opposition to the project, Brian piloted Concorde on its first British flight in 1969. He immediately concluded that Concorde was 'a very precise aircraft to fly' – a sanguine judgement, in view of the fact that he had been required to estimate his height for landing after the failure of the altimeters.

Nor was it the last difficulty with which this aircraft presented him. After its first supersonic flight over land, in 1970, when it travelled at an altitude of 11 miles and faster than a bullet, Brian had to land the plane on three engines because an instrument showed the other to be overheating.

The next year, shortly after receiving the Air League Founders' Medal for outstanding work on the development of the plane, Brian again had to fly on three engines when a metal plate fell off and pieces went through the fourth engine.

Brian was appointed MVO in 1948 and as already mentioned, the CBE in 1970. He received a dozen aviation awards, including, in 1973, the Bluebird Trophy, awarded by friends of Donald Campbell for 'contribution in the realm of high speed'.

In 1972 Brian piloted Concorde on a world tour. That year also produced another first in his life: he met and married my mother.

My mother had already met Lumley Victoria Gertrude Trubshaw – Nine (Welsh for Grandmother) to me – Brian's wonderful mother, some months before at the White-Smiths, and it was she who would later introduce my mother to Brian. I suspect Nine had already considered my mother to be suitable daughter-in-law material after their first meeting. After all, Brian was already 47 and my mother was only five years younger. 'You *must* meet my son darling!', Nine had said enthusiastically to my mother after their first meeting.

Not long after my mother's and Brian's initial meeting, their wedding date was set. Over a month before the wedding my mother dragged me to a shop that, in my opinion, did nothing for street cred, to buy a wedding outfit for me. 'It's *weeks* before the wedding!' I whined. 'We're getting married tomorrow,' my mother said matter-of-factly, 'to avoid the press.' Several days later I was fending off the likes of Janet Street Porter wanting to speak to my mother on the telephone. It appeared my mother was marrying a celebrity.

Brian and my mother had an evening wedding reception a couple of weeks after their wedding. A wedding present had been delivered earlier that day and was purportedly handed over to 'the gardener'. 'The gardener', of course, was none other than Brian himself.

Brian had not enjoyed particularly good health for some years before his death but always remained positive and cheerful as ever. During his last spell in hospital I vividly remember discussing with him the possibility of writing a book about his time spent learning to flying in Phoenix, Arizona, in 1942, when he was 18 – another fairly major time in his life that was skated over in this book. He had started to tell me about some of the fellow students he had trained with, who had been well documented in an album of Brian's full of photographs and press cuttings: all those young men (boys!) at the start of their lives, some of whom did not survive the Second World War in their flying machines and some of whom went on to achieve greatness, aeronautically and in many other different spheres.

We also discussed the possibility of writing about the life of someone who was another great family friend, Sir Geoffrey Tuttle. Uncle Geoffrey, as he liked me to call him, joined the RAF in 1925 on a Short Service Commission – he was still there fifteen years later. He was awarded the DFC before the Second World War. On 31 January 1941 he was taking off in a Hornet Moth from Benson when the engine failed. The aircraft crashed from a height of 50 feet, but fortunately he was not seriously injured. In 1960 he joined the British Aircraft Corporation until 1977, when he became an aerospace consultant until his death in January 1981. Uncle Geoffrey was also an integral part of one of Brian's and my mother's favourite pastimes, the Rugby Internationals at Twickenham, with picnics in the car park first. I always sat next to Uncle Geoffrey at the Internationals, blissfully unaware of his distinguished flying career.

Concorde eventually went into commercial service on 21 January 1976, when British Airways flew to Bahrain and Air France to Rio de Janeiro simultaneously. Two years later, the third test model, Concorde 101, completed what is still the fastest civil transatlantic flight, travelling from Fairford to Bangor, Maine, in 2 hours and 56 minutes.

In 1979 I had my first trip on Concorde, with Jock Lowe at the controls. Brian and my mother had been invited by Goodwood Travel to accompany the flight commemorating the twentieth anniversary of Concorde. Goodwood Travel was at that time providing a Concorde charter service. Brian was offered a fee to entertain the paying passengers with a running commentary from Heathrow to Toulouse. Brian said he would be happy to it without the fee if he could bring his stepdaughter as well. When we arrived in Toulouse, full of champagne, we were treated to an extraordinary presentation of Concorde 201, which had been painted by French University students for its retirement ceremony. We were ushered into a hanger, fully installed with seating for hundreds of people. The 'stage' setting was completely black. Then, as soon as the seats were packed, powerful and dramatic music was pumped around the hangar and the lights came up slowly revealing the static Concorde 201 decked in her psychedelic livery.

After various speeches from French officials, André Turcat and Brian, the three of us were whipped away in a helicopter for lunch in a chateau, and I was thrilled to be sitting on the top table next to Jock. After lunch Brian and Jock each gave one of their incomparable after-dinner speeches to the very happy Goodwood travellers before being flown back to Heathrow in the best aircraft ever built – the Concorde. It was, for me anyway, one of life's rare and special days.

I got to fly on Concorde only one more time, in 1998. Unfortunately my mother was poorly, but Brian and I flew from Heathrow to Farnborough for the launch of his autobiography, with Jock Lowe again at the controls. The flight was full for this short hop, during which time I swapped seats with Raymond Baxter so that he could chat to Brian during the flight. Also among the passengers was Mike Bannister, who was travelling with his family. Brian and I were met on the tarmac by Sutton Publishing, of course, and by Brian's driver Cyril Carroll (the quickest-talking, safest Irish 'racing' chauffeur I have ever met), who, with police outriders in tow, took Brian, myself and Jock Lowe to a celebratory lunch, with other test pilots and well-known greats of the aviation world and their families. It was another rare and special moment.

Brian ended his career as Divisional Director and General Manager of the Filton works of British Aerospace from 1980 to 1986. From 1986 until1993 he was a member of the board of the Civil Aviation Authority, as well as working tirelessly as an aviation consultant.

In 1999 Brian was a passenger as Concorde retraced his first flight to mark its thirtieth birthday. Brian declared that the major difference was that the trip was more luxurious. 'There weren't any seats in the back the first time,' he said.

Brian was devastated when he heard the news of the Paris Air Cash in July 2000 and the subsequent grounding of the service. I was at home in Jersey when I saw him interviewed on television following the crash. I could tell from his face how completely devastated he was. He told the BBC: 'It would be wrong for me to say I was astonished. It was an incident I hoped never would happen, but at the same time one has to be realistic . . . being mixed up with aviation for as long as I have, one knew that one day we could be faced with this situation.' There was no doubt in his mind that Concorde would return to service, as it did some seven months after his death. On 7 November 2001 the fee-paying services were resumed.

I am glad that Brian did not live to see the indignity of Concorde Alpha's fuselage being shipped up the Thames on a barge, the end of another era. It also marked the end of Brian's dream of a successful supersonic passenger airliner – a dream in which he had invested over forty years of his life.

Brian was brave and fearless to the end. When he came home after a long spell in hospital, I knew he was far from well, but he wanted me to get back to work in

Jersey, although I wanted to stay and be useful to both him and my mother.

He was a fighter – an inspiration to all 10-year-olds who see a plane in the air for the first time and know intuitively that they want to fly. But in 1934 the aircraft Brian saw landing on the beach at Pembrey was a far cry from those we see in the sky today. Today's aircraft are flying our skies worldwide because of the work and the expertise of some of the great pioneers of aviation – men like Brian Trubshaw, Test Pilot.

Glossary of Terms and Abbreviations

A & AEE	Aeroplane and Armament Experimental Establishment
ACRC	Air Crew Receiving Centre
ACTUATOR	An electro-mechanical or hydro-mech-anical device fitted to effect some pre-determined linear or rotary movement.
AILERON	Form of simple flap fitted to trailing edge of each wing to provide control in the rolling plane
AILERON BUZZ	High frequency vibration of ailerons
AILERON RODS	Mechanical rods connected from pilot's control to ailerons.
ALPA	Airline Pilots' Association
ANGLE OF ATTACK	Angle between chord line of an aerofoil and the relative airflow
ARB	Air Registration Board
ARTIFICIAL FEEL	Normal aerodynamic control forces provided artificially
ATC	Air Traffic Control
BAe	British Aerospace
BAC	British Aircraft Corporation
BAC (US)	British Aircraft Corporation (USA)
BAFFLE FENCES	Device to hinder or regulate passage of fluid or gas
BEA	British European Airways
BOAC	British Overseas Aircraft Corporation
BUA	British United Airways
BUG/BUGGED	Moveable indicator on dial of

	instru-ment for setting desired condition or indication of it
BWIA	British West Indian Airways
CAA	Civil Aviation Authority
CAC	Concorde Airworthiness Committee
CAD	Commercial Aircraft Division
CDC	Concorde Directing Committee
CDV	Committee Flight Directors
CEV	Centre D'essais en Vol
CFS	Central Flying School
C.G. INDICATOR	Gauge showing centre of gravity
CHAPTER 3 NOISE LEVELS	Definition of present noise standard
DETUNERS	Ground Installation into which aircraft is positioned in order to reduce exhaust noise
DME	Distance measuring equipment
DOI	Department of Industry
DRAG	The component along the drag axis of the resultant force due to relative air flow, the components of thrust being excluded
DRAGON'S TEETH	Set of metal protrusions fitted in front of bomb-bay.
DTI	Department of Trade and Industry
DUTCH ROLL	Rolling and directional yawing motion
EEA	Electronic Engineering Association
FAA	Federal Aviation Agency
FAIRING	A secondary structure added to any part to reduce drag
FLIGHT ENVELOPE	Speed/altitude boundary limits
FTG	Flight Test Group
GCA	Ground Control Approach
GTS	Gas Turbine Starters
HMG	Her Majesty's Government
HSA	Hawker Siddeley Aviation
ILS	Instrument landing system

INTAKE LIP	Leading edge of intake
JCC	Joint Certification Committee
KELVIN	Measurement of temperature
LEADING EDGE DISRUPTERS	Metal protrusion fitted to leading edge of wing
MACH=1.0	= the speed of sound
MANUAL REVERSION	Reversion from hydraulic powered flight controls to manual operation
MCC	Marylebone Cricket Club
MD	Design Mach number limit
MEPU	Term for auxiliary power unit
MOD	Ministry of Defence
N2	High speed compressor speed
NACELLES	Casing surrounding the engine/power plant.
NATS	National Air Traffic System
Og	Weightless condition
OLEO	Shock absorber in the landing gear leg
OTC	Officers Training Corps
OTU	Operation Training Unit
PEAKY WING	Description of particular lift coefficient shape of some wings
PITCH ANGLES	Angles of pitch attitude
PITCH UP	Phenomenon affecting backward swept wings caused by tips stalling before the roots, which generate an increasingly nose-up movement
PNYA	Port of New York and New Jersey Authority
RAE	Royal Aerospace Establishment
RATE OF ROLL	Rate in degrees per second that an aircraft rolls
RATIONAL LANDING TECHNIQUE	Method adopted by ARB for measuring landing distances
RIAT	Royal International Air Tattoo
SAC	(Company started by) Smedley, Allard and Creer

SAS	Stability Augmentation System
SATC	Supersonic Air Transport Committee
SBAC	Society of British Aerospace Constructors
SETP	Society of Experimental Test Pilots
SIA	Singapore International Airlines
SNECMA	Société Nationale d'Etude et de Construction de Moteurs d'Aviation
SPECTACLE INTAKES	Intake shaped like spectacles
SPEED BRAKES	Control surface on wing or rear fuselage that can be extended to reduce speed or increase rate of descent
STICK PUSHER	Device to push stick forward at predeter-mined angle of attack
STICK SHAKER	Electronic motor that shakes pilots' control column
TCA	Trans Canadian Airways
TOA	Japanese Domestic Airline
TRIM CHANGE	Change of control force as speed increases or decreases
TSS	Transport Supersonic Standards
TUFTS	Pieces of wool stuck on aircraft to examine air flow patterns
TWA	Trans World Airlines
V1	Decision speed from which aircraft can either stop or continue within runway distance available
VD	Design diving speed
VHF	Very High Frequency
VOR	VHF Omni Range
VORTEX GENERATORS	Small aerofoils fitted to any part of an aircraft to improve air flow
VR	Rotation speed
VTT	Target threshold speed
WING FENCES	Metal strip fitted across the top of wing chordwise

WING INCIDENCE	Angle of airflow to wing
WOBBLER	Device that wobbles control column through the artificial feel system
WRVS	Women's Royal Voluntary Service
YAW DAMPER	Device fitted to prevent yawing and rolling motion

Index

Index